Computer Modeling and Simulations of COMPLEX BIOLOGICAL SYSTEMS

Edited by
S. SITHARAMA IYENGAR, Ph.D.
IEEE Fellow
Department of Computer Science
Louisiana State University
Baton Rouge, Louisiana

CRC Press
Boca Raton Boston New York Washington London

Acquiring Editor:	Harvey Kane
Project Editor:	Andrea Demby
Marketing Manager:	Becky McEldowney
Cover design:	Denise Craig
PrePress:	Greg Cuciak
Manufacturing:	Carol Royal

Library of Congress Cataloging-in-Publication Data

Computer modeling and simulations of complex biological systems /
edited by S. Sitharama Iyengar.
p. cm.
Includes bibliographical references and index.
ISBN 0-8493-7962-8 (alk. paper)
1. Medicine--Computer simulation. 2. Biology--Computer simulation. I. Iyengar, S. Sitharama.
R853.D37C65 1997
570'.1'13--dc21 97-17685
CIP

International Standard Book Number 0-8493-7962-8
Library of Congress Card Number 97-17685
Printed in the United States of America 1 2 3 4 5 6 7 8 9 0
Printed on acid-free paper

Introduction

S. S. Iyengar

Biological systems are inherently complex information-processing systems. This makes it very difficult and, of course, challenging to model them and to perform computer simulations on them. Physiological complexities of biological systems limit the formulation of hypotheses to explain their behavior and the ability to test such hypotheses. The study of biological systems is concerned with the study and interpretation of biological processes at the molecular level and, more importantly, in terms of the structure and properties of molecules. Indeed, computer technology has an important role to play in such study. The availability of high-performance computers, coupled with mathematical modeling, has contributed to the development of increasingly accurate models of biological systems.

In recent years there has been a tremendous spurt in research and activity in modeling medical data and knowledge representation in the context of understanding physiological complexities, as it is often difficult to predict behavior of biosystems during an experimental investigation. Moreover, biosystems do not behave or act as expected. Their behavior is fundamentally chaotic. However, computer or mathematical models can be helpful in understanding the behavior of complex systems. The process of genetic exchange during replication, the systhesis and regulation of macromolecules such as RNA and proteins, and cancer growth are very difficult to model in a system unless they are actively researched by physiologists or biologists in collaboration with computer scientists and mathematicians. Now, we shall enumerate the computation paradigms embedded in the modeling of these complex systems.

COMPUTATIONAL CHARACTERIZATION

The quest for efficient computational approaches to self-organized data representation in modeling biological systems has undergone a significant evolution in the last few years. Specifically, the application of neural computing concepts to some of the topological properties of biological molecules requires implementation of efficient computer-based strategies with massive processing abilities. For example, the DNA molecule contains an immense amount of information in coded form. This coded information controls the growth, development, and major characteristics of biological systems. Therefore, the control of information stored in a DNA molecule is a major computational problem of research interest for researchers in the discipline of computer science. Therefore, we as computer scientists and biologists must seek to understand the computational potential of this emerging computational paradigm and further explain the fundamental limitations and capabilities of such approaches to modeling complex biological systems. Another interesting parameter for scientists is to understand the embedded importance of computational intelligence in designing complex biological systems.

INTELLIGENT SYSTEMS VS. BIOLOGICAL SYSTEMS

The primary goal of the study of biological systems is to investigate how they behave. This is necessary in order to make recommendations for future development. The focus of modeling biological systems has traditionally been to understand and engineer systems that can address unstructured computational problems (for details see Gulati et al.[2]). One way to approach this problem is to build a computer model for the system through which it becomes possible to predict, control, and optimize the system.[1] However, there are limitations to such approaches.

Intelligent systems, such as expert systems with some embedded reasoning, behave poorly in processing visual or speech information. Contrary to biological systems, they are not adapted to unstructured environments and are unable to learn from experience. Intelligent systems lack such inherent capabilities of biological system as commonsense knowledge and reasoning, structuring knowledge to recognize complex patterns, adaptation, and reorganization to domain-specific queries. Also, intelligent systems fall way behind in taking sensory information and acting upon it, especially when sensors are bombarded by a range of different and competing stimuli. On the other hand, the machinery of biological systems is capable of providing satisfactory solutions to ill-structured problems with remarkable ease and flexibility. A key emphasis underlying any paradigmatic development for unstructured computations in biological systems is focusing on the foundation of biological phenomenon that takes place. This exhibits a spontaneous emergent ability that enables them to self-organize and adapt to their computational structure and function. The central part in emulating biological systems by neural learning lies in the narrowing of differences between organization and structuring of knowledge and between the dynamics of biological neural circuitry and symbolic computational processing paradigms.

OVERVIEW OF THE CHAPTERS

There have been two major trends in the area of modeling complex biological systems:

1. The science of modeling from functionally diverse and distributed data sources.
2. The importance of computing methodologies using genetic algorithms, image processing, neural networks, etc.

In the search for models of complex systems, it is therefore important to concentrate on specific types of tools which are accompanied by large-scale simulation which may be carried out either with digital or analog computers. Toward this objective this book presents a collection of papers on many important components of biological systems.

The first chapter presents paradigms of computational properties of the protein-folding problem, which are very important from the point of view of drug design and creating new polymer-based substances. More specifically, proteins are biomolecules that play a critical role in many biological processes. This protein structure is formed by a sequence of amino acids. Furthermore, proteins have a unique three-dimensional structure which determines the biological functionality of the protein. In summary, Chapter 1 presents a unique way of representing these computational properties of protein-folding problems.

Genetic algorithms are quite robust in solving many computational search problems in biological systems. These algorithms are superior to traditional optimization methods such as conjugate gradient descent. The first chapter explores this method in an overview fashion. The second chapter presents a distributed genetic algorithm for a conformational search process in biological systems using PARAM, based on SuperSparc, often referred as PARAM 9000/SS. The novelty in this chapter is in exploring a very efficient method of solving a search problem using genetic algorithms for this computationally intensive problem.

Chapter 3 describes another method of genetic coding for image segmentation problems. The formulation and implementation of a randomized search approach to segment an image using genetic algorithems is presented in this chapter. Images from two different scenes are segmented using this genetic approach to illustrate the applicability of such systems. Chapter 4 presents a control theory and computer-modeling approach to a cancer problem. In particular, the chapter presents a unique approach to model a characteristic of cancer, that is, uncontrolled cell proliferation in biological systems. It also addresses a scenario of how these proliferations of normal and abnormal cells can be regulated at the organ, molecular, and cellular level.

During the last 10 years, HIV has created a lot of research interest in the medical community. The understanding of the functioning and structure of this virus has become a paramount concern. Toward this objective, Chapter 5 presents a model of HIV infection and its blockers, from complex to simple. More specifically, this chapter provides an interesting chronological order to the progressive simplification of a model of HIV infection. Chapter 6 discusses a mathematical modeling and understanding of virus infections. The chapter discusses models of HIV infection kinetics in tissue cultures and in humans. Of utmost importance for life processes on Earth are the cell cycles. Toward this end, Chapter 7 presents mathematical models of cell cycles.

There is an increasing interest in the development of population growth models as they become a fundamental part of most modern ecological systems. Chapter 8 outlines the recent advances to the stochastic population growth modeling.

Life as we understand it is essentially an information-processing system governed by a number of biological and natural processes. There are many apporaches to modeling such processes. Toward this goal, Chapter 9 provides a comprehensive review of the reconstructability analysis approach in the modeling of biological systems. Unlike classical methods, this approach provides an unbiased representation of data and their components. There are numerous biological neural network systems that will integrate sensory information for the purpose of viewing a biological system as a whole. Chapter 10 presents a computer simulation of bimodal neurons and networks integrating infrared and visual stimuli.

The chapters essentially integrate methodologies across a variety of disciplines for the sake of modeling complex biological processes and systems. High-performance computing technology and simulation are essential to understand these complex computational processes in biological systems, as is demonstrated in these chapters.

The book is written primarily for the benefit of graduate students and professionals who may be interested in employing interdisciplinary or multidisciplinary approaches to solve complex problems pertaining to biological systems. Others who will find it valuable are biologists, chemists, engineers, research physicians, mathematicians, and computer scientists. The editor wishes to acknowledge the assistance and encouragement of several individuals, including the contributors to this book. Every chapter in this book reflects entirely the author's opinion which may or may not agree with the view of the editor.

REFERENCES

1. Iyengar, S. S., Ed., *Computer Modeling of Complex Biological Systems,* CRC Press, Boca Raton, FL, 1984.
2. Gulati, S., Barhen, J., and Iyengar, S. S., Neurocomputing formalisms for computational learning and machine intelligence, *Adv. Comput.,* 33, 1991.

Editor

S. S. Iyengar, Ph.D., is Chairman and Professor of the Department of Computer Science at Louisiana State University and an IEEE Fellow. Since receiving his Ph.D. from Mississippi State University in 1974, he has served as a principal investigator on research projects funded by the Office of Naval Research, National Science Foundation, U.S. Army Research Office, Department of Energy, Naval Research Laboratory, NASA Jet Propulsion Laboratory, and various agencies of the state of Louisiana.

Dr. Iyengar has made significant contributions to the areas of high performance image processing algorithms, sensor fusion, parallel models of computation, robot motion planning, and computer vision modeling systems. He has published several books and monographs in addition to over 250 research articles.

Dr. Iyengar has served as guest editor for *IEEE Computer, IEEE Transactions of Data Knowledge Engineering, IEEE Transactions on Systems, Man, and Cybernetics, IEEE Transactions on Software Engineering,* the *Journal of Computers and Electrical Engineering,* and the *Journal of the Franklin Institute.* In addition to being an IEEE Fellow, he is also a distinguished visitor of the IEEE and a member of the New York Academy of Sciences. In 1996, he was awarded the LSU Distinguished Faculty Award and the Tiger Athletic Foundation Teaching Award. He has supervised over 27 Ph.D. dissertations and over 40 M.S. projects and theses while at LSU.

CONTRIBUTORS

Dimiter S. Dimitrov
Laboratory of Experimental and Computational Biology
National Cancer Institute
National Institutes for Health
Frederick, Maryland

Werner Düchting
Department of Electrical Engineering and Computer Science
University of Siegen
Siegen, Germany

Thomas Ginsberg
Department of Electrical Engineering and Computer Science
University of Siegen
Siegen, Germany

Terry L. Huntsberger
Intelligent Systems Laboratory
Department of Computer Science
University of South Carolina
Columbia, South Carolina

S. S. Iyengar
Department of Computer Science
Louisiana State University
Baton Rouge, Louisiana

B. Jones
Department of Computer Science
Louisiana State University
Baton Rouge, Louisiana

Thomas R. Kiffe
Department of Mathematics
Texas A & M University
College Station, Texas

Kurt W. Kohn
National Cancer Institute
National Institutes for Health
Bethesda, Maryland

A. S. Kolaskar
Bioinformatics Centre
University of Pune
Pune, India

James H. Matis
Department of Statistics
Texas A & M University
College Station, Texas

John R. Rose
Intelligent Systems Laboratory
Department of Computer Science
University of South Carolina
Columbia, South Carolina

Guna S. Seetharaman
Center for Advanced Computer Studies
University of Southwestern Louisiana
Lafayette, Louisiana

John L. Spouge
National Center for Biotechnology Information
National Institutes for Health
Bethesda, Maryland

V. Sundararajan
Centre for Development of Advanced Computing
Pune University Campus
Pune, India

S. K. Trivedi
Computer Science Department
Southern University
Baton Rouge, Louisiana

Waldemar Ulmer
Max-Planck-Institute
Munich, Germany

John M. Zachary
Department of Computer Science
Louisiana State University
Baton Rouge, Lousiana

Dedicated to my parents
and wife, Manorama Iyengar

Table of Contents

1 Computational Aspects of the Protein-Folding Problem

*John M. Zachary and S. S. Iyengar**

CONTENTS

1.1 INTRODUCTION

Proteins are biomolecules which play a crucial role in many biological processes. Formed by a sequence of amino acids, a protein has a unique three-dimensional structure which determines the biological functionality of the protein. An important open question is "how does the amnio acid sequence determine the three-dimensional structure of the protein?" and the answer will influence novel drug design, understanding of illness, and the creation of new polymer substances. This chapter explains the computational aspects of the protein-folding problem from the perspective of computer science. There are three main parts. Section 1.2 is an introduction to protein science, Section 1.3 presents the computational challenges of the protein folding

* The authors may be contacted by E-mail at {zachary,iyengar}@bit.csc.lsu,edu. This work was partially funded by a DOE grant from the Oak Ridge National Laboratory.

0-8493-7962-8/98/$0.00+$.50

TABLE 1.1
Twenty Naturally Occurring Amino Acids Found in Proteins

Alanine	Leucine
Arginine	Lysine
Asparagine	Methionine
Aspartic acid	Phenylalanine
Cysteine	Proline
Glutamine	Serine
Glutamic acid	Threonine
Glycine	Tryptophan
Histidine	Tyrosine
Isoleucine	Valine

problem, and Section 1.4 focuses on global optimization techniques, a crucially important computational aspect of the protein-folding problem.

1.2 BASIC CONCEPTS OF PROTEINS

1.2.1 INTRODUCTION

As previously mentioned, proteins are biopolymers formed by a linear sequence of amino acids with a three-dimensional structure uniquely determined by the sequence. In the cell, DNA molecules issue coded instructions through messenger RNA (ribonucleic acid) which then pass the code to transfer RNA. It is transfer RNA working in conjuction with ribosomes that actually generates the proteins required by the cell. Four RNA nucleotides, adenine, guanine, cytosine, and uracil, work together with amino acids to produce these proteins. Amino acids may also be referred to as *monomers*. There are 20 naturally occurring amino acids (Table 1.1).

The molecular structure of an unionized amino acid molecule is given in Figure 1.1 where R is called a *side chain* and is the distinguishing chemical component among the amino acids in Table 1.1. Given that the proteins of all living organisms are constructed from these 20 amino acids, proteins are indeed complex and remarkable biomolecules.

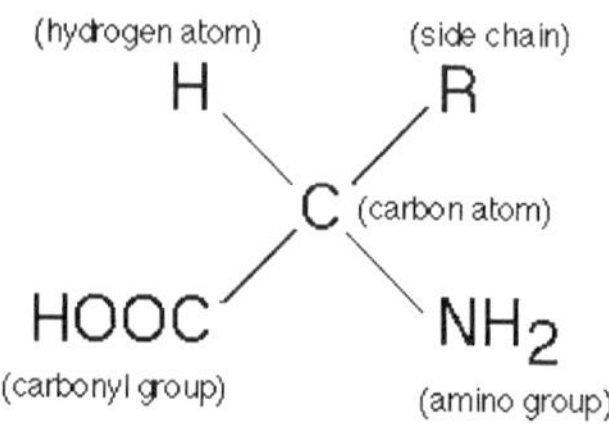

FIGURE 1.1 The abstract structure of an unionized amino acid.

The importance of proteins to biological organisms is elucidated by their roles in the following functions[7]:

1. *Enzymatic catalysis* in chemical reactions spanning hydration to chromosome replication;
2. *Transport and storage* of molecules and ions such as O_2 by hemoglobin in the bloodstream and myoglobin in muscle tissue;
3. *Structured support* by collagen, a fibrous protein found in skin, bone, and tendons;
4. *Muscular contraction* leading to coordinated motion by two filamentous proteins;
5. *Immune protection* by the abililty of antibodies to recognize and collude with foreign bodies such as virii and bacteria;
6. *Nerve impulse origination and transmission* made possible by receptor proteins such as acetylcholine at nerve synapses and rhodopsin in retinal cells; and
7. *Growth and differentiation control* by genetic material and growth factor hormones.

Proteins are classified into three main types depending on the structure and function of the given monomer sequence. *Fibrous proteins* play a structural role in the body, *membranic proteins* participate in storage and transport functions, and *globular proteins* are primary components in enzymatic catalysis (the initiation and promotion of chemical reactions). The attention in the remainder of this chapter is on proteins of the latter type.

1.2.2 Protein Structure

The native folded state of proteins is normally spherically compact and unique. The core of the protein consists of those amino acids which are repulsed by water and are referred to as *hydrophobic* monomers. The outer shell of the protein contains those polar amino acids which easily form hydrogenic bonds with the surrounding aqueous solution, otherwise referred to as *hydrophilic*. It is this latter characteristic of interaction with the protein's environment combined with the unique structure of the protein that defines its functionality.

As previously mentioned, proteins are composed of amino acids arranged in a linear sequence. The composition of the sequence is the *primary structure* of the protein. Amino acids are joined to one another by peptide bonds which form peptide groups which are themselves planar and rigid. The degree of rotational freedom in the protein which leads to the well-defined three-dimensional structure of the protein is found on either side of the peptide bond (Figure 1.2). The rotation of the bond between the two carbon atoms is denoted as Ψ and the rotation around the bond between the nitrogen atom of the peptide group and the carbon atom of the non-peptide group is denoted Φ. The essence of the protein-folding problem corresponds to values of Ψ and Φ at each peptide group along the amino acid sequence. Ramachandran plots are contour diagrams which show allowed regions of values for

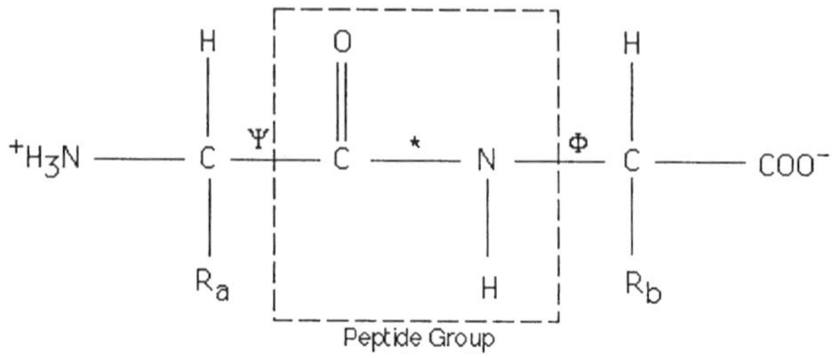

FIGURE 1.2 Peptide groups and degrees of rotational freedom in inter-amino acid bonds.

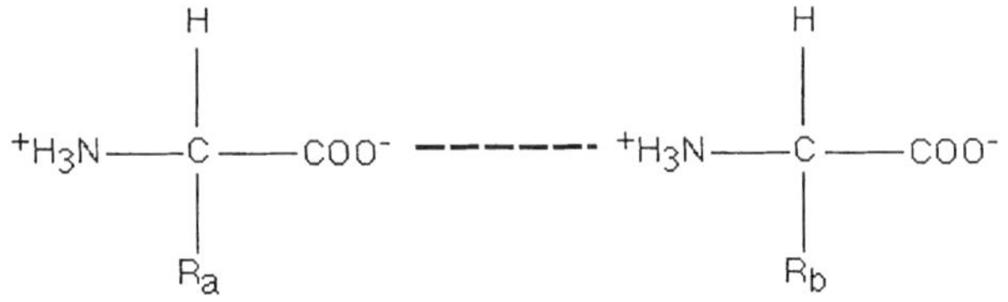

FIGURE 1.3 Example of an electrostatic bond between two oppositely charged atoms in two amino acids.

certain chain classes. Further insight into conformation structure is revealed by research showing that certain amino acid classes exhibit different frequencies of secondary structure. Stryer[7] provides further detail on these two subjects.

It was discovered by Linus Pauling and Robert Corey in 1951 that polymer chains of amino acids can fold into two periodic structures called *α-helices* and *β-sheets* which together form the *secondary structure* of the protein. These structures arise by adjacent and local interactions of amino acids in the sequence. The importance of this discovery is that precise knowledge of component properties of polymer sequences can lead to a prediction of the conformation of the amino acid sequence.[7] The *tertiary structure* of a protein is a result of nonlocal interactions between amino acids far apart in the monomer sequence and contributes to the binary division between polar and nonpolar amino acids in the spherical shape of a protein.

1.2.3 Physical Driving Forces in Protein Conformations

There are three types of bonds in reversible interactions between biomolecules. Each differs in strength, geometry, and interaction with water.[7]

Electrostatic bonds occur between two molecules of opposite charges. Figure 1.3 is an example of this type of bond. Electrostatic bonds, also referred to as *ionic bonds,* are described by Coloumb's law

$$F = \frac{q_+ q_-}{r^2 D}$$

TABLE 1.2
Hydrogen Bonding Behavior of 11 Amino Acids

Hydrogen Bond Donors Only	Hydrogen Bond Donors and Acceptors	pH-Dependent Donors and/or Acceptors
Tryptophan	Asparagine	Lysine
Arginine	Glutamine	Aspartic Acid
	Serine	Glutamic Acid
	Threonine	Tyrosine
		Histadine

where q_+ is the charge of the positive group, q_- is the charge of the negative group, r is the measure of the distance between the two charges, and D is called the dielectric constant and is dependent on the surrounding medium. Electrostatic bonds are very strong over small distances.

Hydrogen bonds are formed between charged or uncharged molecules in which a single hydrogen atom is shared by two atoms, such as oxygen and nitrogen. It is important to note that hydrogen bonds are highly directional from the hydrogen donor atom to the hydrogen receptor atom. Stryer[7] categorizes the side chains of 11 of the 20 naturally occurring amino acids into three classes based on their behavior in forming hydrogen bonds (Table 1.2).

Hydrogen bonds play a particularly important role in protein conformation. In 1936, Linus Pauling and Alfred Mirsky discovered that hydrogen bonding as a result of protein conformation leads to the secondary structure of proteins, α-helices and β-sheets.[2] Hydrogen bonds are formed between nonconsecutive amino acid groups and may be local or nonlocal.

While hydrogen bonding plays a critical role in protein conformation, it does not exclusively determine the three-dimensional structure of proteins. In an aqueous solvent, an unfolded amino acid sequence will contain monomers which will tend to form hydrogen bonds with water molecules before bonds with other amino acids are formed. Hence, some other force must also assist in driving the monomer sequence into a conformed state. The other force responsible for this is *hydrophobicity*. As mentioned earlier, proteins have a structure in which the core consists of mainly hydrophobic amino acids while the surface amino acids are mostly hydrophilic. Thus, before the amino acid sequence can begin to form α-helices and β-sheets, the hydrophobic amino acids begin a collusion process to form the protein core and bring the sequence together so that the hydrogen bonding between amino acids can occur. Therefore, hydrophobicity together with local and nonlocal hydrogen bond interactions are the major forces driving proteins to a native conformed state. This viewpoint is expounded by Chan and Dill.[2,3] Another viewpoint suggests that other proteins are also necessary for conformation *in vivo*.[4]

van der Waals bonds occur between atoms which are within a few angstroms of one another. Although it is weaker than the other two bond forces previously described and is nonspecific in the choice of bond partners, van der Waals bonds play an important role in protein conformation. A large number of van der Waals bonds which occur simultaneously lead to a geometric specificity in the molecule. That is, a large number of these types of bonds can only occur if the molecule is conformed in a small space, such as are proteins. In a native state protein, the amino acid is tightly packed and hydrogen side chains form van der Waals bonds within the protein. Thus, van der Waals bonds contribute to the molecular stability of a native state protein structure, but they do not actively participate to a significant degree in the conformation process.

Computationally, these forces of folding could be simulated on supercomputers by techniques (from molecular dynamics) except that the number of amino acids in a typical protein constrains the simulation to within a few nanseconds of the folding dynamics process. The total timescale for the protein to conform from an unfolded state to a native folded state is on the order of 10 to 10^3 s.[7] Clearly, new computational techniques are required to push the envelope of protein-folding dynamics simulation further along the timescale. The next section focuses on the computational aspects of the folding process from the principles of the forces previously described.

1.3 COMPUTATIONAL CHALLENGES IN PROTEIN FOLDING

Physical systems do not reach a state of thermodynamic equilibrium by exhaustively searching all possible states until the state corresponding to a minimum energy is found. Instead, some process guides the system to the minimum energy state. This concept is applicable to protein folding; proteins cannot realistically try all possible conformations in search of the native state structure in the time it normally takes a protein to reach the native state, usually on the order of 10 to 10^3 s. We can define a free-energy surface dependent on the N degrees of rotational freedom of the amino acid sequence,[2]

$$\varepsilon_f = f\left(\Psi_1, \Phi_2, \Psi_3, \ldots, \Psi_{N-1}, \Phi_N\right)$$

Even for small proteins with a number of atoms on the order of $\approx 10^5$ atoms, ε_f is a high dimensional space. Newton's laws of motion might be used to model the amino acid interaction energies, but due to the harmonic oscillations of the atoms participating in bonding, numerical integration techniques require extremely small time steps (on the order of 10^{-15}). This makes the direct approach of searching ε_f using molecular dynamics force field simulations virtually intractable at this time.

However, since we describe the situation in which hydrogen bonding and hydrophobicity are the major factors in protein conformation and van der Waals bonds contribute to structure stability when the protein structure is compact, we should use a technique to search ε_f which maximizes the effects of hydrogen bonding and hydrophobicity. In effect, what we are proposing is that an objective function which achieves the maximization can be used in a global optimization scheme to search the free-energy surface for the optimal value corresponding to the native conformation of the protein. In the next section, we will describe global optimization techniques in general and provide some suggestions of their applicability to the protein-folding problem.

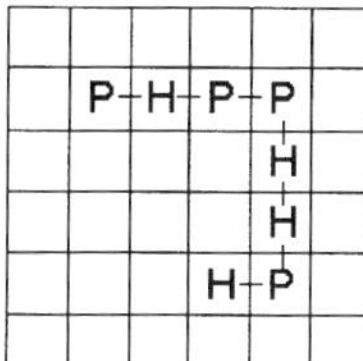

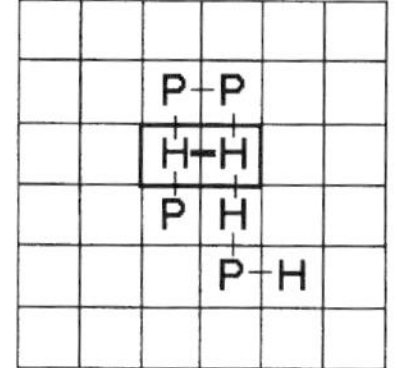

FIGURE 1.4 HP lattice model of Chan and Dill showing two amino acid sequences. The first figure is a free-form sequence without adjacent H monomers. The second chain shows an example of a folded sequence with two adjacent H monomers and the bond between them (shown in a heavy line).

We will now describe the approach of Chan and Dill[2] called the *HP lattice model* (Figure 1.4). This model assumes that hydrophobicity is the primary driving force in protein conformation and specifies proteins as a chain of symbols, H and P, representing the tendency of the amino acid to be hydrophobic or hydrophilic, respectively. This chain is then embedded into a two- or three-dimensional lattice as a non-self-intersecting walk. If two H monomers are spatially adjacent in the lattice but no bond exists between them, then that interaction is given an energy value $\epsilon < 0$; all other interaction energies are valued at 0. The energy of the conformation is then the sum of all interaction energies, and the minimum energy arrangement corresponds to the native state of the protein. Interestingly, the same secondary structure that arises from protein folding, α-helices and β-sheets, was also observed in the lattice models. The lattice models also exhibit mutational plasticity, or the ability of a protein to retain its structure given small sequence alterations. For small amino acid sequences, this process can be enumerated over all possible chain walks, but it is not a computationally efficient technique in general. Hence, it is useful to consider global optimization strategies in generalizing the utility of HP lattice models to arbitrary amino acid sequence lengths. For further details of the HP lattice model, the reader is referred to References 2 and 3.

1.4 SURVEY OF GLOBAL OPTIMIZATION TECHNIQUES

1.4.1 PRELIMINARIES

Nonlinear optimization problems are prevalent across much of mathematics and science. The general task is to find the minimum value of some function called the *objective function*. In protein folding, we have hinted at the components that might necessarily need to be embodied in an objective function. As we indicated, to search the energy surface of the conformation space corresponding to an amino acid sequence, it is desirable to maximize local and nonlocal hydrogen bonding between sequence monomers as well as the effects of hydrophobicity between such amino acids and the surrounding medium. In this way, we are forcing the sequence into a globular structure which consists of a hydrophobic core and secondary structure of α-helices and β-sheets. We can also consider that we may want to maximize the van der Waals forces in the structure. This component may be used to compare local minima and avoid entrapment in such local basins, as will be discussed. However, we must also assure that the protein structure is physically viable, i.e., has no self-intersections. Thus, geometrical and topological constraints must be incorporated into the objective function. Finally, a robust representation for the domain of the objective function will depend on the degrees of rotational freedom in the sequence.

Generally speaking, a good search method uses two mechanisms to find an optimal value:

1. A local search method, such as gradient descent;
2. A method to avoid entrapment in local minima basins.

There are two types of optimization methods, local and global. The classification of a method generally depends on which mechanism the method chooses to emphasize (note that without a loss of generality, the terms "optimization" and "minimization" are interchangeable).

1.4.1.1 Local Optimization Methods

Methods in this category are characterized by fast local optimization strategies, but they suffer from entrapment in the local minima basins they converge to. Examples of this type of optimization method are conjugate gradient descent and Newton's method. There are three pathologies that cause these methods to fail:

1. When the surface is flat in the local neighborhood being searched, $\nabla f = 0$. Thus, these methods will halt.
2. A wide range of gradient values will cause particular problems. Small gradient values will cause the method to converge slowly while large gradient values may cause the method to overshoot extrema or diverge. Newton's method is a good example of this latter phenomenon.
3. If the surface of the objective function is rugged, then the local optimization methods tend to converge to local minima values close to the point at which the search began.

1.4.1.2 Global Optimization Methods

Optimization methods with the goal of finding the globally optimal value place much of their computational effort on avoiding entrapment in local minima basins. Gradient descent is generally still used in finding a local minima, but then, by vastly different methods, the search process shifts to finding a point with a lower functional value than the current local minima.

Probabilistic global optimization methods rely on probability theory to escape local minima basins. Examples are clustering algorithms, genetic algorithms, simulated annealing, and stochastic methods. This type of method is generally more computationally expensive than the deterministic method to be discussed, and they lack a well-defined stopping criteria. However, these types of methods, particularly genetic algorithms, have experienced a recent surge in interest. Deterministic methods, on the other hand, use deterministic heuristics to escape from local minima basins. We present two optimization methods to demonstrate these concepts. The first method is the popular notion of genetic algorithms. The second method is a novel deterministic algorithm called TRUST based on the concepts of subenergy tunneling and terminal repeller dynamics.

1.4.2 Genetic Search Algorithms

Genetic algorithms (GAs) are based on the Darwinian mechanism of natural selection and genetic operations. They combine "survival of the fittest" among a pool of string structures with a structured randomized information exchange to create a self-improving device, analogous to the process of natural selection. Each new generation inherits the best string features of the previous generation to achieve better performance.

The basic computational element of GAs is an encoding of an instance of the problem into a string. The string is the genetic material with each position, or gene, assuming some value, or allele. The objective function to be minimized serves as the gauge of fitness of the string in the population of strings, or gene pool. In the protein-folding problem, a natural choice for the string encoding is to represent each degree of rotational freedom in the amino acid sequence as a single allele. However, this encoding can lead to long strings which, as will be discussed, can cause the genetic algorithm approach to become inefficient.

The basic psuedocode for a typical GA is as follows

1. Randomly generate an initial population of M strings
2. While the algorithm has not converged
   - 2a. compute the fitness value of each of the M strings
   - 2b. if the deviation of the fitness values from the mean is sufficiently small then the algorithm has converged
   - 2c. otherwise,
     - (1) calculate the selection probabilty for each string p = e/E where e is the individual score and E is the total score over all strings
     - (2) generate a new population using the selection, crossover, and mutation operations

At each stage of this meta-algorithm, the number of offsprings of a string S in the next generation is proportional to the success of string S in the current population. This success is dependent on certain patterns in the string called *schema* which contribute to favorable objective function values. In general, these patterns are propagated to future string generations. This concept of schema is the algorithmic analogue of the natural concept of survival of the fittest.

There are three basic genetic operators used in most GAs.

1. *Crossover* is the exchange of allele values between two strings. It may occur at one or several points. A simple example of crossover is given two strings, ABCDE and FGHIJ, crossover at the third gene position would result in two new strings ABHIJ and FGCDE. The point(s) where

the crossover occurs is arbitrary. Some strategies pick the crossover point at random while others have some heuristic to choose the best position adaptively.

2. *Inversion* is an internal form of crossover. Given the string ABCDE, an inversion operation might result in the new string ADCBE. Again, the point and scope of the inversion is an arbitrary decision.
3. *Mutation* is a background operator which, unlike the previous two operations, does not necessarily occur at each step of the algorithm. Mutation involves making a random change in allele values, such as ABCDE to ABSDE. The purpose of mutation is to theoretically ensure that the entire search space is reachable. Thus, it serves in some part to escape local minima entrapment and begin new searches in a nonlocal neighborhood of the current search which might be functionally more optimal than the current search neighborhood.

For further details of these operations, the reader is referred to Goldberg.[5]

GAs are quite robust when compared with traditional optimization methods, such as conjugate gradient descent. This robustness is a result of the difference between GAs and other traditional methods. First, GAs manipulate data which is an encoding of the parameter set into genes. Traditional methods, on the other hand, work with the parameter set directly. This implies that GAs can be used on problems with nonderivative, discontinuous, and unimodal domains as long as these domains can be encoded. The second way in which GAs differ from traditional methods is that GAs search from a population of domain values, the string population. Most traditional techniques begin searching from a single point and use derivative information to guide the search to new points. Thus, GAs are an inherently parallel process which simultaneously search all possible direction via the three genetic operations. In its most basic form, GAs use only the objective function value to guide its search. As mentioned in the previous paragraph, traditional methods usually depend on auxiliary information about the objective function, such as derivatives, to guide the search process. Because less information is needed to minimize the objective function, a wider range of problems are solvable using GAs. The transition rules of GAs are probabilistic in nature, and, thus, with structured random search, GAs guide the search process to regions of the search space with the highest probability of improvement.

However, GAs have certain drawbacks. First, the time complexity of GAs is usually much greater than traditional methods because of the fact that multiple search points are being processed. By decreasing the size of the gene pool, the genetic algorithm loses some degree of its inherent parallelism. Thus, there is usually some trade-off between the amount of time for searching and flexibility of the gene pool. Another disadvantage of GAs is they lack a well-defined stopping criterion. This is due to their probabilistic nature. As will be seen in the next section, other deterministic methods do have stopping conditions which contribute to their degree of reliability. Also, problems which depend on a high degree of accuracy may suffer from long string lengths when GAs are used to solve them. The result is a long convergence time. Finally, the problem to be solved must be encodable into a string, and that encoding must be resilient to crossover, inversion, and mutation operations. However, the strings of some problems, for example the Traveling Salesman Problem, may become meaningless if one of these operations occur (this example is left as an exercise for the reader). Thus, methods to encode the encodings must be used, adding to the complexity of the algorithm.

1.4.3 The TRUST Algorithm

Barhen and Protopopescu[1] have proposed a deterministic optimization method called *terminal repeller unconstrained subenergy tunneling* (TRUST) which avoids local minima entrapment and avoids an exhaustive global search for the optimal value. Compared with other methods, its performance in terms of function evaluations is quite favorable.

The objective function to be optimized, $f(\bar{x}) : \boldsymbol{R}^n \to \boldsymbol{R}$ is assumed to be semicontinuous with a finite number of discontinuities. Other assumptions are

1. The domain $\mathfrak{D}$ of f is compact and connected. A result of this is that f is finite on $\mathfrak{D}$ given that f is semicontinuous and $\mathfrak{D}$ is compact.
2. Every local minima of f is twice differentiable, and

   a. $$\frac{\partial f(\bar{x}_{lm})}{\partial \bar{x}} = 0, \text{ and}$$

   b. $$\bar{x}^T \frac{\partial^2 f(\bar{x}_{lm})}{\partial \bar{x}^2} \bar{x} \geq 0, \quad \forall \bar{x} \in \boldsymbol{R}^n$$

The TRUST method is based on the idea of deterministic tunneling to avoid local minima entrapment. The main idea of TRUST is to transform the objective function f into a new function $E(x,x^*)$ with similar extrema properties but where a current local minima of f, x^*, is a global maximum of $E(x,x^*)$. A functional value of f less than $f(x^*)$ is then found by applying gradient descent to find a local minima of $E(x,x^*)$ with starting point x^*. This is accomplished by applying two concepts to this process: subenergy tunneling and terminal repellers. The general algorithm is as follows:

1. Use gradient descent to find a local minima of f at x^*.
2. Transform f into the *virtual objective function*

   $$\mathrm{E}(\mathrm{x}, \mathrm{x}^*) = \mathrm{E}_{\mathrm{sub}}(\mathrm{x}, \mathrm{x}^*) + E_{\mathrm{rep}}(\mathrm{x}, \mathrm{x}^*)$$

   where

   $$E_{\mathrm{sub}}(x, x^*) = \log\left(\frac{1}{1 + e^{-(f(x) - f(x^*)) + a}}\right)$$

   is a nonlinear monotonic transformation of f called the *subenergy tunneling term* (see Figure 1.5) and

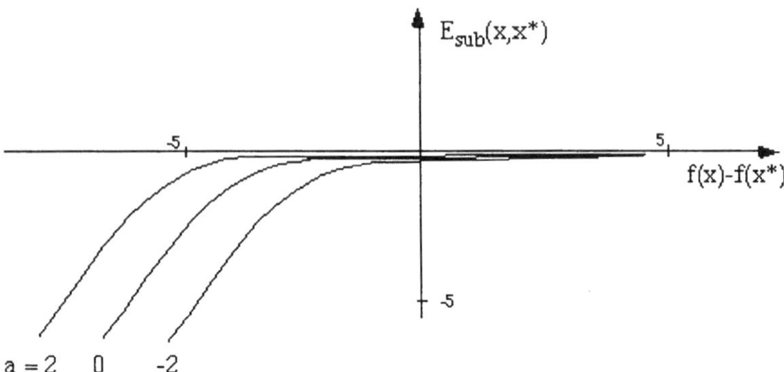

FIGURE 1.5 Monotonic behavior of E_{sub}. For values $f(x) - f(x^*) < 0$, the behavior of E_{sub} is unchanged; for $f(x) - f(x^*) \geq 0$, E_{sub} rapidly converges to 0.

$$E_{rep}(x, x^*) = -\frac{3}{4}\rho(x - x^*)^{4/3} u(f(x) - f(x^*))$$

is the *terminal repeller term* (u is the Heaviside function).

3. Use gradient descent to find a local minima of E_{sub} starting at x^*. This yields the following dynamical system

$$\dot{x} = -\frac{\partial E}{\partial x} = -\frac{\partial f}{\partial x}\frac{1}{1 + e^{-(f(x)-f(x^*))+a}} + \rho(x - x^*)^{1} 3u(f(x) - f(x^*))$$

An equilibrium state of $\dot{x}$ is a local minimum of $E(x,x^*)$ which corresponds to a local minimum of f with a lower function value than $f(x^*)$.

The subenergy tunneling term, $E_{sub}(x,x^*)$, is used to isolate the curve segments of f which are functional less than the value $f(x^*)$. The difference term $f(x) - f(x^*)$ offsets f such that x^* tangentially intersects the x-axis. Thus, $E_{sub}(x,x^*)$ is equivalently 0 when $f(x) - f(x^*) \geq 0$ and identical to $f(x) - f(x^*)$ when $f(x) - f(x^*) < 0$.

The repeller term $E_{rep}(x,x^*)$ guides the gradient descent search in the third step from the globally maximum value at x^* to a local minimum of $E(x,x^*)$. (For more details on terminal repellers, see Barhen and Protopopescu.[1])

Figure 1.6 graphically portrays this process for a one-dimensional function. In the one-dimensional case, the TRUST method is guaranteed to converge to the global minimum of f. In the multidimensional case, while no formal proof yet exists for convergence, strategies exist to balance between the guaranteed accuracy and computational cost, one of which is to reduce the multidimensional problem into a one-dimensional problem via hyperspiral embedding. Another strategy is to augment the repeller term weights based on the gradient behavior of f to guide the search in Step 3 to the closest highest ridge value of $E(x,x^*)$. The reader is referred to Barhen and Protopopescu[1] for details of these strategies.

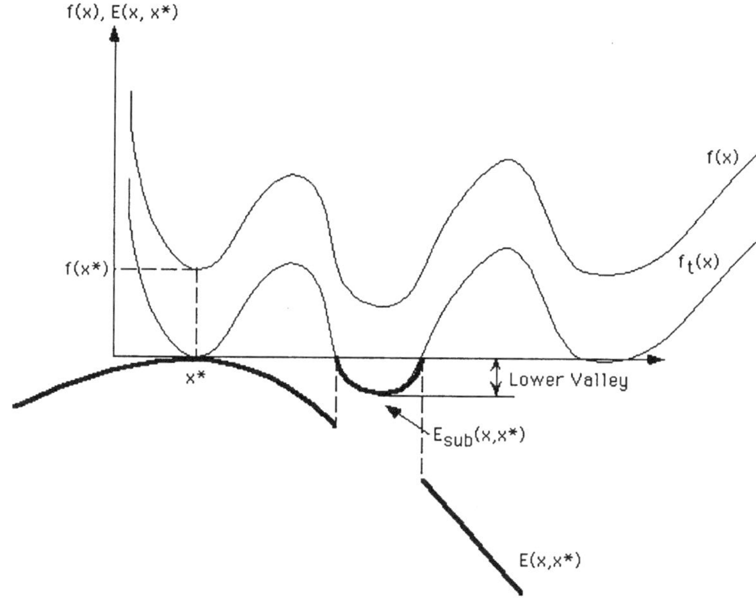

FIGURE 1.6 The TRUST method for a one-dimensional optimization problem.

The advantages of TRUST over other global optimization methods are

1. TRUST exhibits good computational performance with respect to the number of functional evaluations.
2. Convergence is guaranteed in the one-dimensional case with strong assurances for convergence in the multidimensional case.
3. There exists a definite stopping conditioin for the TRUST method.
4. TRUST is amenable to parallelization.

1.5 CONCLUDING REMARKS

This chapter has set out to describe the process of protein folding from the point of view of the computational strategies that are required to solve the problem. We have focused on the global optimization of an energy surface determined by the degrees of rotational freedom in a protein and the molecular forces which affect them (hydrogen bonding, hydrophobicity). We then gave a description of two global optimization techniques that attempt to solve the problem from different philosophies. GAs are based on the Darwinian concept of natural evolution and are probabilitic in nature. TRUST global optimization is a deterministic approach which is based on two concepts: subenergy tunneling and terminal repellers. In order to solve the protein-folding problem, advances in global optimization techniques are crucially required.

REFERENCES

1. Barhen, J. and Protopopescu, V., Generalized TRUST algorithms for global optimization, in *State of the Art in Global Optimization,* Floudas and Pardalos, Eds., Kluwer Academic Publishers, New York, 1996, 163–180.
2. Chan, H. S. and Dill, K. A., The protein folding problem, *Phys. Today,* 2, 24–32, 1993.
3. Dill, K. A., Bromberg, S., Yue, K., Fiebig, K. M., Yee, D. P., Thomas, P. D., and Chan, H. S., Principles of protein folding – a perspective from simple exact models, *Protein Sci.,* 4, 561–602, 1995.
4. Gething, M. J. and Sambrook, J., Protein folding in the cell, *Nature,* 355, 33–45, 1992.
5. Goldberg, D. E., *Genetic Algorithms in Search, Optimisation, and Machine Learning,* Addison-Wesley, New York, 1989.
6. Press, W. H., Teukolsky, S. A., Vetterling, W. T., and Flannery B. P., *Numerical Recipes in C: The Art of Scientific Computing,* Cambridge University Press, U.K., 1992.
7. Stryer, L., *Biochemistry,* W. H. Freeman Co., New York, 1988.

2 Distributed Genetic Algorithms on PARAM for Conformational Search

V. Sundararajan and A. S. Kolaskar

CONTENTS

Abstract — A model has been implemented using genetic algorithms to predict the structure of a polypeptide chain. The algorithm is based on the principle of evolution, and it improves the solution of the posed problem by genetic operation crossovers and mutations. Dihedral angles are taken as the basic variables for the structure of the molecules, and genetic operations are carried over on a population of binary strings of (ϕ, ψ) angles. Sequential, data parallel, and distributed model versions have been developed. Better efficiency in optimization has been observed in a distributed model for the case of an octapeptide.

0-8493-7962-8/98/$0.00+$.50

2.1 INTRODUCTION

Biomolecules are heterogeneous polymers having complex structures. The structure of these molecules decides their function. Hence, it is important to know the structure of the biomolecules. Experimentally, X-ray diffraction and nuclear magnetic resonance (NMR) spectroscopy techniques are used to determine the structure. However, the molecules are required to be synthesized in large quantity to grow single crystals of sufficiently large size for X-ray diffraction or high resolution NMR studies in solution. If neither is possible, theoretical prediction is the only alternative. So, an ability to predict the structure would go a long way in the applications of medicine and biotechnology, especially, in drug design. Structure prediction or conformational search is normally treated as an optimization of energy of the molecule. The calculus-based minimization procedures are good only at locating a local minimum on the energy surface. On the other hand, while dynamic quenching is computationally demanding, one is not sure of reaching the global minimum energy structure. Recently, randomized techniques like simulated annealing (SA) and genetic algorithms (GA) have been widely used in the search of the global minimum. SA is based on the search by carrying out a random process of disturbing the conformations and adopting a slow and step-by-step *cooling* procedure to arrive at the global minimum energy conformation. On the other hand, GA is based on the search by random selection of conformations and performing the genetic operations, specifically crossover and mutation, to arrive at the global minimum. GA is not only efficient algorithmically, but also quite amenable to parallelization. The present work exploits both of these aspects.

There have been earlier attempts based on GA to predict the conformation of biopolymer-protein and polypeptide.[2,9-11] Sun[9] used a reduced representation model to predict the folded structure of proteins using GA. That model assumes the molecular structure of each protein reduced to its backbone atoms (with ideal fixed bond lengths and valence angles) and each side chain approximated by a single virtual united atom.

Schulze-Kremer[8] optimized the structure of the Crambin molecule using GA by fixing the side chains as well as the bond lengths and bond angles. The search was carried out in the torsion angle space. He used the best set of torsion angles picked up from databases in the initial population. The genetic operations are performed on real coding rather than on binary coding. Schulze-Kremer[8] and Sun[9] restrict the (ϕ, ψ) angles to the allowed region in the Ramachandran plot or pick up the most probable angles from the database whenever the angles go out of range.

In a recent paper, Hermann and Suhai[2] applied GA to dipeptide analogues. They were able to locate a better minimum than that by the other methods.

In the present work we have modeled an oligopeptide and have restricted the variation only in the torsion angles. We have also assumed the side chains to be fixed. We evolved the set of conformations in torsion angle space using GA for (1) dipeptide and (2) homo-octapeptide. We present these results, as well as our experiences on the simulations. We have parallelized our code to exploit the parallelism in GA as well as the power of the machine PARAM 9000 developed at the Center for Development of Advanced Computing (C-DAC). Detailed aspects of implementation of GA and our experiments on its performance are also presented.

The rest of this chapter is organized, such that the second section describes GA in brief and gives implementation details of GA. Parallel versions of GA are discussed in Section 2.3. The fourth section enumerates sample applications and their results. The last section draws the conclusion and also summarizes the present work.

2.2 ALGORITHM AND IMPLEMENTATION

GA is based on the theory of evolution of biological systems.[1,3] In short, GA evolves the solutions for the posed problem through a fitness function of individuals in a population. The population of conformations stored in memory undergoes reproduction and variations by processes similar to mutation and crossover. These iterated operations would lead to better and better conformations and at the end are expected to provide improved solutions for the posed problem. The global optimum is the result of evolution of a population of individuals simultaneously. Thus, each evolutionary process can be studied in parallel mode. This could be further exploited by explicit parallelization, which is reported here.

A typical implementation of GA would require the following: (1) a string representation, (2) definition of the objective function, (3) a suitable selection procedure, and (4) the meaning of genetic operators like crossover and mutation. This section elaborates these aspects to obtain minimum energy conformation of oligopeptides, polypeptides, and proteins using the GA approach.

2.2.1 Representation

There are two ways of representing conformation of biopolymer chains, specifically, external and internal coordinates. In the present work, we use the internal coordinates — bond lengths, bond angles, and dihedral angles. In that we fix the bond lengths and bond angles and vary only the dihedral angles (ϕ, ψ). One can also vary side chain dihedral angles χ, and such a variation is straightforward to incorporate in the present implementation. We use binary strings arranged in the order (ϕ_1, ψ_1), (ϕ_2, ψ_2), …, (ϕ_n, ψ_n) for an n-peptide (each angle represented by a nine-bit binary string). Parameters (ϕ, ψ) are folded back to the range $-\pi$ to $+\pi$ whenever they go out of the range.

2.2.2 Objective Function

We use the energy of the molecule as the objective function given by

$$E = E_{nb} + E_{es} + E_{tor} \tag{2.1}$$

where E_{nb} is the nonbond energy, E_{es} is the electrostatic energy, and E_{tor} is the torsion energy due to dihedral angle variations. The lower the energy, the better the fitness.

We used the parameters listed by Ramachandran and Sasisekharan[4] for the nonbond and torsion angle interactions without hydrogen bond. No penalty is incorporated in the implementation of algorithm carried out by Schulze-Kremer.[8]

2.2.3 Selection and Reproduction

Biomolecular conformations with lower energies are selected and replicated. The selection process is done by using the tournament selection method. Here, we randomly shuffle the list of conformations. Pick up the first two and choose the one with lower energy. Then, pick up the next two, and so on, to get half of the new population. Repeat the above process to choose the next half of the new population. This ensures two chances for each conformation and hence retains the best and deletes the worst.

2.2.4 Crossover

Two randomly selected binary strings representing sets of (ϕ, ψ) angles in the new population, called *parents*, exchange their *bits* in the *string* by probability p_c. We considered a one-point crossover, where the exchange of the binary strings takes place from a randomly selected location to the end of the string. We did not consider two-point and uniform crossover in this report.

2.2.5 Mutation

Mutation is carried out by probability p_m on any of the bits in the strings of the new population. It helps in taking the system out of the local minimum. In other words, mutation ensures that any part of the conformational space is accessible.

2.2.6 Simulation Schedule

A typical simulation would require one to go through the iterations of selection, crossover, and mutation on a population of the size of the order of string length (100 in the case of an octapeptide). So, one needs to do experiments on the parameters, specifically, on the probabilities of crossover and mutation. The probability of crossover is varied in the range 0.5 to 0.7 since this range has been tested and found fruitful on a number of applications.[1] The occurrence of mutation is generally very rare; hence, we use the range from 0.001 to 0.01. Having fixed these, the only other parameter required is the number of iterations. We did some experiments on this and found that the number of iterations between 100 and 200 is ideally suited for an octapeptide. This may vary from case to case. That amounts to spanning about 20,000 points in a search space of 2^{126} points. The simulations are performed a number of times to ensure statistically that what is achieved at the end is an optimal solution and not just a local minimum. Two tests have been used; the first is to check whether or not the final set of torsion angles lies within the allowed region and the second is to check for no short contacts between the atoms.[4]

2.2.7 Results of Sequential GA

2.2.7.1 Dipeptide

This is the most important of all the case studies, because this forms the building block of the rest of the investigations.

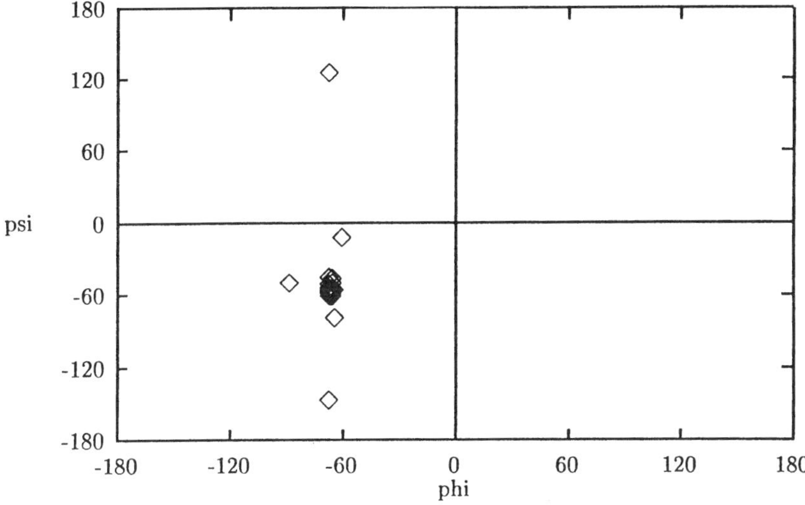

FIGURE 2.1 The population of dipeptide conformations in the (ϕ, ψ) space at the end of the simulation for a dipeptide.

We used a population of 50, randomly generated the conformations, and went through about 20 iterations. As the iterations proceeded the distribution of torsion angles was focused to second and third quadrants (Figure 2.1) by probabilistically driven selection and genetic operations. These angles lie well within the allowed regions of the Ramachandran plot. This clearly demonstrates that GA is capable of evolving from a random set of (ϕ, ψ) angles and gives confidence that GAs would work for predicting conformation of a polypeptide and protein. Similar results have been reported by Hermann and Suhai.[2]

2.2.7.2 Homo-Octapeptide – Ala_8

This is another test for GA to check whether or not it can evolve a helical turn from a random set of torsion angles. We used Ala_8 to check this as it is known that a chain of *Ala* forms a helical structure.

We proceeded in the following manner to get an optimal structure for this case. We performed about five simulations each of about 200 iterations. Each time the randomseed is changed to have different starting populations for the GAs. The following observations have been made based on these simulations.

The final population of conformations represented by torsion angles for different runs has most of the angles lying within the allowed region.

The performance plot is shown for a typical run in Figure 2.2. In this, we plot energy vs. iterations. The top curve represents the total average energy of the molecule in the whole run. The lower curve is the global minimum energy for the molecule. From this curve it is evident that the global minimum obtained does not change beyond 120 iterations. In addition, we check that there are no short contacts

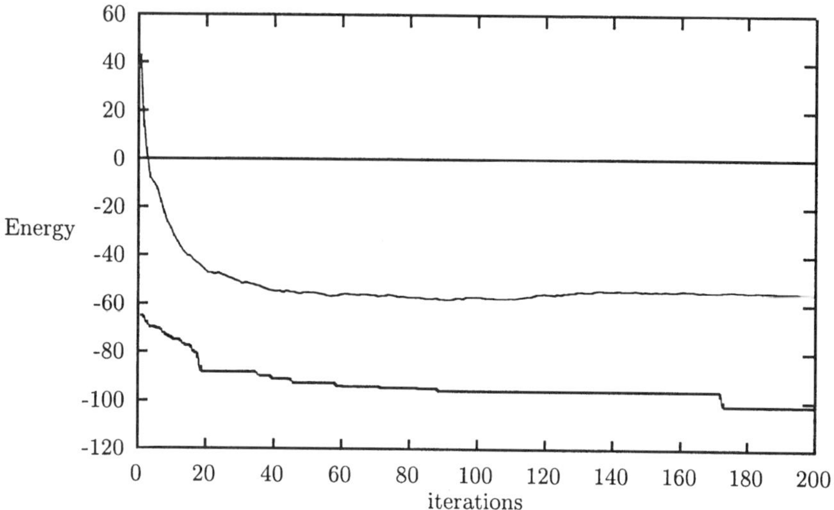

FIGURE 2.2 Performance plot for the sequential GA in the case of homo-octapeptide. The thin curve gives the average energies, and the thick curve gives the minima picked up as the stimulation proceeds.

among the atoms in the molecule. These are the two criteria that we impose for locating the global minimum.

Octapeptide has been used to check the implementation of the parallel version.

2.3 PARALLEL IMPLEMENTATIONS

GA is one of the ideal examples of natural parallelism. In the iterative process of GAs the most time-consuming part of the calculation is the individual fitnesses. First, this has been implemented in a data parallel version that distributes the calculation of fitnesses to different processors. Further, we have implemented a distributed model, also known as a *migration* model, which deviates from conventional GAs in using concurrent evolution akin to nature. Before we describe these models and present the results, a brief outline is given about the PARAM machine that has been used in the project.

PARAM is a series of supercomputers developed by C-DAC implementing open-frame architecture. The first in the series is PARAM 9000/SS based on Sun super sparc (SS) processors. The architecture is built around a multistage interconnect network which used a packet-switching wormhole router as a basic element. Each switch is capable of establishing 32 simultaneous nonblocking connections to provide a sustainable bandwidth of 320 Mb/s. The communication links have built-in support for standard environments like message-passing interface (MPI) and a parallel virtual machine (PVM). The architecture allows the parallel machine to be viewed as an ensemble of independent workstations or as a massively parallel processing (MPP) system connected by a high-bandwidth network. The present simulations are carried out using a computer program written in FORTRAN 77 with

TABLE 2.1
Performance on PARAM 9000/SS

No. of Processors	Time in secs	Speedup	Efficiency
1	956.2	1.0	1.0
2	511.9	1.9	0.95
4	256.2	3.7	0.93
8	130.0	7.4	0.93
12	90.9	10.5	0.87
16	71.5	13.4	0.84

an MPI standard. The code is portable to any other parallel machine or cluster of machines.

2.3.1 Data Parallel Model

As mentioned above, the calculation of fitnesses at the end of each generation loop is totally independent. This is a simple data parallel model and is implemented by a master–slave approach embedded in a single program. The communication pattern is a simple broadcast of parameters by a master required by each slave for fitness calculation. After the calculation of the fitnesses on each processor (both master and slave), the master collects the results from each slave. Synchronization of the tasks is effected after each iteration. The required communication calls are taken care of by two subroutines — *broadcast* and *collect*. The former broadcasts the ϕ,ψ angles and related data to all the processors, and the latter is used to calculate the fitnesses on each processor and collect them back at the master. The master process proceeds to the next generation of selection, crossover and mutation which does not take even one 100th of the time taken for fitness calculation. This accounts for a minimal idling time of the slave processors and results into a near-linear speedup for the case of an octapeptide. Table 2.1 gives the timings and speedup up to 16 processor PARAM 9000/SS, amounting to about 83% efficiency. This also includes the waiting time of the slave processors for the master to complete the initialization part. Moreover, the current implementation uses MPI which in turn uses transfer control protocol/Internet protocol (TCP/IP). Implementation with low-latency protocols and microkernels is expected to enhance the efficiency further, at least by 5 to 10%.

2.3.2 Distributed GA Model

In the distributed GA or migration model, several islands of subpopulations evolve simultaneously, i.e., GAs on a set of processors are distributed, initiated, and left to themselves for a cycle of a fixed number of iterations. Periodically, say, after every n iterations, each processor would send an x% of conformations (solutions) to its neighbor (on the right). The communication pattern and topology in this case is a ring. After this process of migration, each processor would replace x% of its worst

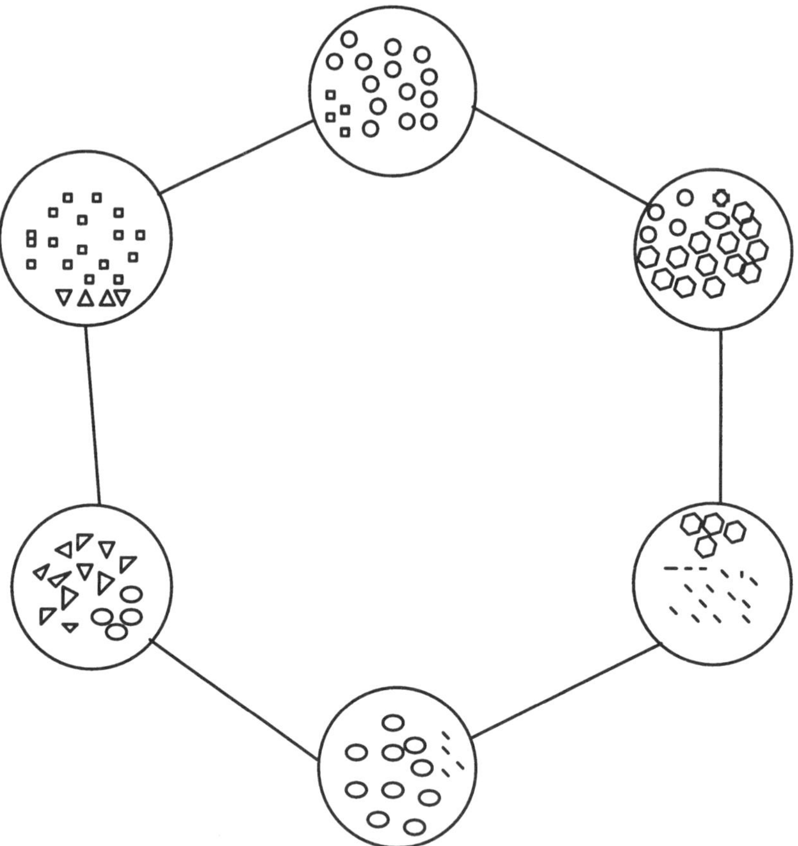

FIGURE 2.3 Migration model showing six processors running GAs and also showing the transfer of the few individuals across to the next neighbor.

solutions with the best ones sent by its left neighbor and proceed with another cycle of n steps before the next event of migration. This process is repeated until there is no further improvement in the optimization.

In Figure 2.3 a set of six circles is shown with different geometric objects belonging to them. In order to illustrate the migration process, a few of these objects are shown to be present in the right neighbor of each circle. This kind of distributed GA has several advantages.

1. A single run in parallel is equivalent to a set of GA runs on a sequential processor. Normally, GA is run with different starting points, and the solution that is the best out of all these runs is picked up. So, there is a definite reduction in computation time.
2. It brings in more diversity because of the additional operation of migration which may help in realizing the optimal result faster than sequential GA. Thus, efficiency is enhanced as well as the search space explored more efficiently by avoiding possible stagnation.

3. There are a number of ways by which one can exploit the distributed GA by the variation of the percentage of migration, the number of migrants, as well as the pattern of migration. We tried one or more variation — instead of the best ones, randomly selected ones are exchanged.
4. Communication overhead is minimal; hence, parallelization is more interesting.

This model has been used by Gorges-Schleuter et al.[6] and also discussed by Dorigo and Maniezzo.[5] Results of the simulations using this model are given in the next section.

Mansour and Fox[7] have compared their implementations of parallel SA, parallel neural network, and parallel GA on an optimization problem of data allocation. Although slow, parallel GA produced the best results. They used the shifting balance theory for distributed GA in which weakest individuals are placed with the fittest migrants from the neighborhood. In the current implementation the major differences are in the process of selection limited to a GA island and in migration, which is limited only to a particular neighbor.[10] For the case of an octapeptide, this has been found to be sufficient. More variations would be implemented in order to maintain diversity in the case study for larger molecules.

2.4 HOMO-OCTAPEPTIDE — Ala_8

This is a test for GA to check whether or not it can evolve a helical turn for an octapeptide of alanine from a random set of torsion angles. A simulation using the migration model has been performed with six processors. This experiment requires two quantities — migration frequency n and the percentage of migrants x. We fixed n to be 50 and x to be 10, i.e., every 50th iteration the best solutions are replicated and migrated to the GA island to the right. We performed two trials, one with migration and the other without migration. For both cases, the results of performance have been plotted in Figure 2.4.

In this, the dotted curves are the results without migration and the continuous curves are those with the migration operator *on* with the above-mentioned migration frequency and percentage of migrants. The plot is essentially of the minimum-most energies in each iteration picked up during the whole simulation. It is evident that there is a distinct improvement in minimization in the case with migration, seen by the drops in the continuous curves just after the migration occurred, i.e., at 50, 100, and 150 iteration numbers. In fact, all six concurrent GAs show improvement compared with their counterparts without migration having the same starting initial sets. The minimum energy has improved from 2 to 10%, which is crucial for large molecules. The three-dimensional structure of the best solution is shown in Figure 2.5 as a ball-and-stick model with a ribbon strand as a guide to show the helical turn very clearly.

Also, the final population of conformations represented by torsion angles for different runs has most of the angles lying within the allowed region. This is one of the important observations, confirming that one can start from a set of random torsion angles and arrive at the proper conformation of a biomolecule.

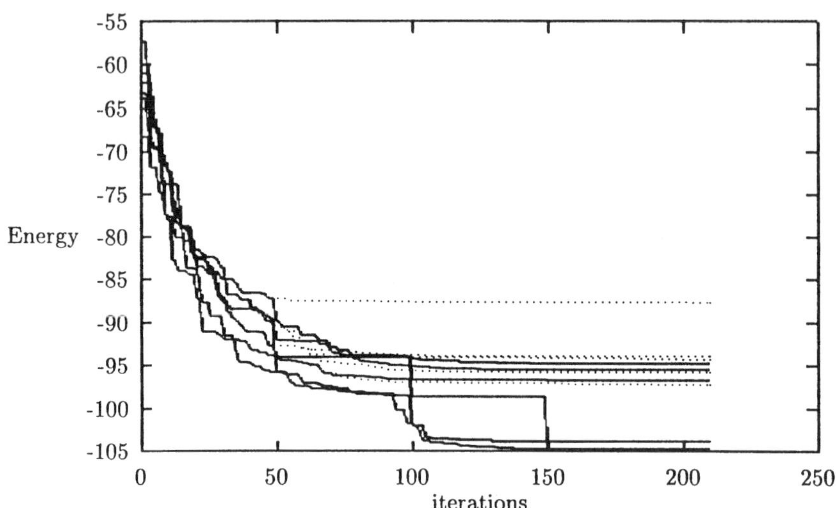

FIGURE 2.4 Performance plot in the migration model. The dotted curves are for the case of GA islands working concurrently without migration. The continuous curves are the case where there is a 10% migration after every 50 iterations.

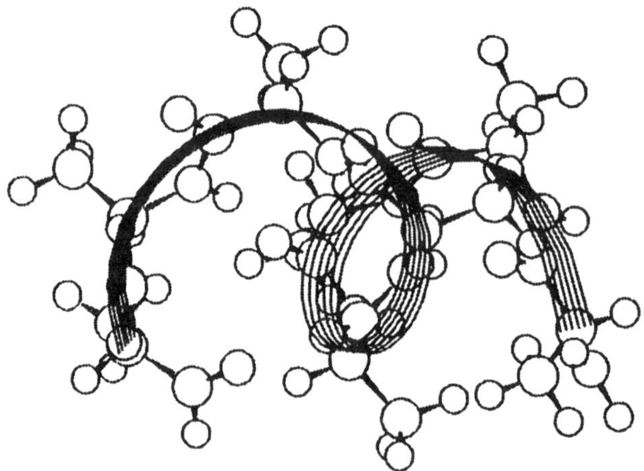

FIGURE 2.5 Three-dimensional structure of the homo-octapeptide. The best structure obtained by the migration model. Ribbon strands are shown as a guide to the eye for helical turns. (Generated using Insight II of Biosym Inc.)

2.5 CONCLUSION

A GA-based method has been presented to optimize the structure of a dipeptide and an octapeptide. The algorithm gives encouraging results in the sense that we are able to start from a population of random solutions and evolve the global minimum through the processes of selection and genetic operations, specifically, crossover and mutation. We used a binary string to represent the (ϕ, ψ) angles of the individuals in the population. The code was parallelized first to distribute the calculation of fitness values to different processors of the parallel machine PARAM. This gives a near-linear speedup. Algorithmic parallelization has been implemented using the *migration* model and found to give performance better than the sequential one by about 2 to 10% which may be crucial for larger molecules. Work on the prediction 3-D structure of larger biomolecules-proteins is in progress. At the same time, efforts are also made to reduce the communication overheads using low-latency high bandwidth protocols.

ACKNOWLEDGMENT

The authors would like to thank Dr. Kalyanmoy Deb and Dr. Devapriya Choudhury for many fruitful and exciting discussions.

REFERENCES

1. Goldberg, D. E., *Genetic Algorithms in Search, Optimization, and Machine Learning,* Addison-Wesley, New York, 1989.
2. Hermann, F. and Suhai, S., Energy minimization of peptide analogues using genetic algorithms, *J. Comput. Chem.,* 16, 1434, 1995.
3. Holland, J. H., *Adaptation in Natural and Artificial Systems,* 2nd ed., The University of Michigan Press, Ann Arbor, 1989.
4. Ramachandran, G. N. and Sasieskharan, V., Conformation of polypeptides and proteins, *Adv. Protein Chem.,* 23, 283, 1968.
5. Dorigo, M. and Maniezzo, V., in *Parallel Genetic Algorithms,* J. Stender, Ed., IOS Press, Amsterdam, 1993, 5.
6. Gorges-Schleuter, M., Muhlenbein, H., and Kramer, O., Evolution algorithms in combinatorial optimization, *Parallel Comp.,* 7, 65, 1988.
7. Mansour, N. and Fox, G. C., Parallel physical optimization algorithms for allocating data to multicomputer nodes, *J. Supercomp.,* 8, 53, 1994.
8. Schultze-Kremer, in *Parallel Genetic Algorithms,* J. Stender, Ed., IOS Press, Amsterdam, 1993, 129.
9. Sun, S., Reduced representation model of protein structure prediction: statistical potential and genetic algorithms, *Protein Sci.,* 32, 762, 1993.
10. Sundararajan, V. and Kolaskar, A. S., Parallel genetic algorithms on PARAM for conformation of biopolymers, in *Proceedings of the Third International Conference on High Performance Computing, IEEE Computer Society,* Trivandram, Dec. 19–22, 1996, 22.
11. Pedersen, J. T. and Moult, J., Genetic algorithms for protein structure prediction, *Curr. Opin. Struct. Biol.,* 6, 227, 1996.

3 A Genetic Coding Structure and Algorithms for Image Segmentation

Guna S. Seetharaman

CONTENTS

Abstract — The formulation and implementation of a randomized search approach to segment an image, using genetic algorithms, is presented in this chapter. A state-space representation of a partially segmented image introduced by Brice and Fennema[2] is suitably adopted into binary strings, in which the dominant substrings are easily modeled as chromosomes. Also, operations, such as crossover and mutation, are easily abstracted. A modified crossover operator using boundary interaction and region adjacency graph (BIRAG) has been adopted to improve the performance. Also, a simplified mutation operator called a "switch" has been devised using the BIRAG. In particular, the specific data structure used in this scheme also facilitates a means for accommodating pixel-level feedback and model-based bias for model-based segmentation of

0-8493-7962-8/98/$0.00+$.50

images. Images from two different scenes are segmented using this approach to illustrate the applicability of such systems.

Segmentation of an image refers to partitioning the image into several subimages,[1] such that each subimage provides the basis for a *logical entity* or a *geometric entity* in the final interpretation of the original image. In this context, *geometric primitives,* such as points, lines, arcs, and regions, act as the building blocks, and *logical entities* are often described using these building blocks.

The purpose of segmentation varies vastly with applications. In most biological systems its purpose is to separate objects of interest from the background. In machine perception of images, it amounts to the extraction of geometric and logical entities from images — an important task for applications involving visual inspection. The existence of a unique and truly optimal segmentation, and its sensitivity to the sampling process, is yet to be established.

Almost any ambulatory biological creature, endowed with a vision faculty to sense its dynamic environment, exhibits a capacity to perform this task effortlessly. The speed and resolution at which these systems categorize their surroundings are very crucial to their very survival. Complex computational methods unique to biological systems have an advantage over sequentially ordered, causally progressive computations (performed by machines).

Experimental evidence suggests that humans fully use our past experience (with similar images) to decide the resolution, speed, and approach to use to segment a given image. At the same time, we maintain a balance between trying new approaches and following the past. This is innately required in order not to be deceived by optical illusions and cognitive illusions. Image segmentation using genetic algorithms captures the flavor of complex computational models of human and biological systems.

3.1 INTRODUCTION

Segmentation plays an important role in low-level processing of any vision system. A formal and more-precise definition of segmentation is found in Pavlidis.[1] Image segmentation attempts to partition the given picture into disjoint sets of pixels in such a way that a predefined uniformity criterion is satisfied over each set individually, and not so over the union of any two sets that are spatially adjacent.

The flow of control and data in segmentation algorithms follow one of four paradigms: macromodel-driven algorithms; semantic-model-driven algorithms; statistics-driven algorithms; and random search algorithms. The macromodel-driven paradigm includes edge detection–based techniques; facet model–based techniques, and almost all template–based methods. Statistics–based methods include clustering, region growing, and region splitting, etc.

The edge-detection and contour-tracing methods follow a *macroscopic model*–driven approach. In this paradigm, a local measure indicating the disparity between two adjacent pixels is computed by a set of edge operators. Edge operators are, by design, sensitive to spatial discontinuities in image intensity. Convolving these operators to an image produces a vector quantity at every pixel describing the strength and the direction of image discontinuities called an edge at that pixel. These edges are then grouped into meaningful contours which will divide the given image into final segments. The method, essentially, is a bottom-up procedure. The edge operators inherently involve differentiation; therefore, they are extremely sensitive to noise. So parametric models of local surfaces have been used to overcome noise and the orientation-sensitive nature of edge detectors.

A second popular method is the region-growing approach, which follows the *statistical measure*–driven approach. First, a set of homogeneous and spatially connected pixels called a seed region is identified. The seed region is to be iteratively grown as big as it can get by assigning at each iteration an adjacent (neighboring) pixel to that region if their union satisfies certain uniformity constraints. The iteration terminates when there are no more pixels to be assigned to the region. Also, the segmentation is considered complete when every pixel has been assigned a region.

A dual approach to region growing is the split-and-merge procedure[3] which follows a hierarchical control structure. One starts with the entire image as a candidate segment — instead of a number of seed regions. A candidate segment, by definition, is a connected set of pixels which is not necessarily homogeneous, and every candidate segment which is homogeneous forms the basis for a region to be extracted later. The procedure, then, iteratively keeps splitting each nonhomogeneous candidate into several new candidates until all the candidate regions are homogeneous. Then, one could merge these homogeneous candidate blocks into bigger groups of homogeneous regions. Both the split and the merge steps are thus necessary in order to facilitate regions of arbitrary shapes and sizes.

Region-based techniques using the classical state-space approach was first applied in Reference 2 and extended in Reference 4. This approach regards the initial image as a discrete state and each pixel as a separate region. Transitions in states take place when a boundary is inserted or removed between any two regions. The problem is thus transformed into searching for allowable changes in the state space to find the best partitioning, where the nonuniformity measure is the cost function to be minimized. Often the heuristic cost functions used in these processes employ some scene-dependent knowledge, hence, the lack of generality.

The problem is at least as hard as a well-known graph problem[5] called *partitioning a graph*. Given a graph, the graph-partitioning problem involves a search for a partition of its nodes such that a given cost function is minimized. This problem is *NP*-complete. Whereas the graph problem partitioning tries to optimize a well-defined cost function, most segmentation algorithms do not. The major bottleneck in designing a universal segmentation program is associated to its subjective and application-dependent nature. It is implicit that each segment has to represent a logical entity in the scene. Then, the underlying metrics to be satisfied varies vastly with applications. Various factors contribute to the difficulties in segmenting an image; some are application specific and most are not.

Given the fact that the problem is NP-complete, it is worth examining randomized search algorithms that are being applied to similar problems. The research presented in this chapter extends the classical state-space-based approach developed in Reference 2 and later expanded in Reference 6, and regards image segmentation as a combinatorial optimization[2] problem. The segmented image is considered a discrete state, in a space of $O(2^{n^2})$ states. Transitions in states take place when an edge is inserted or removed between two regions. The problem then is to search for the optimal state, for which a cost function is minimized, by a series of allowable changes called *valid state transitions*.

Any standard approach such as "hill climbing" will not work, unless delayed gratification is taken into account. Simple search procedures are likely to be trapped at local minima. The insertion of a single edge is not likely to change the segmentation, as significantly as the removal of an edge does. A major problem in these methods is associated with designing a suitable cost function that is to be optimized. The nature of the space is such that there exist many to one representation between the points in the state space and their corresponding segmentation. Using the terminology of hill-climbing techniques, the cost functions are said to be largely flat, with a large number of plateaus in the search space.

In particular, the state-space representation lends itself to a simplified representation of segmentation, or partition, by a binary string of length $2n^2$ bits, thereby a practical candidate for genetic algorithm. For example, when all the bits are turned on, it corresponds to the instance where every pixel is a region. From an algorithmic design standpoint, this model of representing a segmentation captures the inherent nature of the problem. In the simplest end of the scheme, a given segmentation may be improved by selecting a few boundary pixels of a segment and reassigning them to a neighboring segment. The semantic structure of the segmented image can be kept almost the same while improving the accuracy in locating the boundary pixels. On the other extreme, a more global view may demand a radically different semantic structure in the segmented image. There is a practical need to have both capabilities. Design of a model-based vision system would be significantly enhanced by such segmentation schemes. Psychovisual evidence indicates that such an evidence-triggered regrouping of pixels does occur in human vision systems. Human perception of occluded contours and subjective contours,[7] in some sense, may be incorporated in machine vision, through such strategies. A prototype of this approach to dynamic segmentation, called segment inspect-and-refine algorithms (SIRA),[8] attempts to do that. A number of carefully chosen data structures, capable of performing set-union, set-membership, set-reshuffling operations are necessary to go through the many alternate solutions rapidly. SIRA employs such data structures.

The rest of the chapter describes various factors that influence the genetic algorithm–based approach to image segmentation. The basic representation of candidate segmentation is explained in the next section. An $O(\log n)$ algorithm of generating n – *choice* discrete variable with a given bias is also explained. Then, exact values of the bias are derived from the given image. Given a $2n^2$-bit binary string, a simple labeling algorithm can be used to group the pixel in $O(n^2)$ time, for an $n \times n$ image. How the initial population of candidate segmentation may be generated using these techniques is described. The definition of gate functions based on strongly correlated bits is given, describing its use for crossover operation. A simple, boundary interaction, region adjacency graph (BIRAG) representation of segmented images also permits identification of and discrimination among various possible gate functions that may be used for crossover operations. Gate functions are essentially a set of bit positions that represent all the bits that must be considered simultaneously.

The rest of this chapter contains a detailed description of suitable data structures, appropriate methods to generate the initial population, and mutation as well as crossover operations, etc. Two images are segmented to illustrate the applicability of the concept.

3.2 BASIC MODELS OUTLINED

3.2.1 Problem Definition

The input to a segmentation procedure is an image $f(i,j)$, defined over a two-dimensional discrete space G of finite size. In general, G is a rectangular grid, $G = \{(x,y) \mid 0 \leq x,y \leq N\}$. The segmented image will be defined such that $s(x,y) = k$, if the pixel (x,y) belongs to the segment k.

The goal of a segmentation algorithm is to divide the given picture into many connected sets of points, such that

1. A certain set of uniformity is satisfied over these connected sets individually, and not so over the union of any two spatially adjacent connected sets.
2. In general, each segment *connected set of points* will correspond to a logical entity present in the scene represented by the picture at hand.

A connected component, by definition, is a set $S \in G$ of pixels, such that for every pair of pixels x_b, $x_e \in S$ there exists at least one path x_b, …, x_e starting from x_b and ending at x_e, consisting only of those pixels that are in S. One exception to that rule occurs when both these points are simply connected. Then, there are no intermediate pixels, hence an empty path $\emptyset$, which conceptually is a subset of S. Thus, the segment labeled k, say S_k, may represent a set:

$$S_k = \{(x,y) \mid s(x,y) = k\}$$

Then, a segmented image, as whole, is represented by S_1, S_2, S_3, …, S_K, such that

1. $S_1 \cup S_1 \cup S_2 \cup S_3 \cdots \text{U } S_K = X \in G$
2. The predicate $P_1(S_i) = true$, $\forall_i$, $1 \leq i \leq K$
3. $S_i \cap S_j = \emptyset$ if $i \neq j$.
4. The predicate $P_1(S_i \cup S_j) = false$; $\forall_{i,j}\ i \neq j$.

The purpose of the predicate P_1 is to ensure that each partition is uniform and homogeneous in itself. The predicate P_1 is defined, such that $P_1(X_i) \equiv C(X_i) \wedge (Q_1(X_i) < T)$, where (1) $C(X_i)$ ensures that X_i is a *connected set*; (2) $Q(X_i)$ a quantitative measure of the smoothness of the function $f(x,y)$ over the points $(x,y) \in X_i$.

$$Q(X_i) = \frac{\sigma_i^2}{\sigma_0^2}$$

$$\sigma_i^2 = \text{Var}\big(\phi(x, y) \,\big|\, (x, y) \in X_i\big)$$

$$\sigma_0^2 = \text{Var}\big(\phi(x, y) \,\big|\, (x, y) \in X_0\big)$$

Also, the purpose of the predicate P_2 is to ensure that large regions do not get cut into several small regions.

The predicate P_2 is defined, such that it refers to the second-order relations on two *adjacent* regions.

Case adjacent (X_i, X_j)

True : $P_2 : \neg P_1(X_i \cup X_j) \wedge \big(Q(X_i \cup X_j) > T\big)$

False : $P_2 = true$

endcase

Note that the *adjacent* (X_i, X_j) function is defined as (1) *true* iff $\exists\, \boldsymbol{x}_1 = (x1, y1)$, $\boldsymbol{x}_1 \in X_i$ and $\exists\, \boldsymbol{x}_2 = (x2, y2)$, $x_2 \in X_j$, such that $\boldsymbol{x}_1$ and $\boldsymbol{x}_2$ are 4-connected neighbors.

Then finding the right segmentation becomes an optimization problem that minimizes a certain cost function. The objective then is to find a partitioning, $X = X_1 \cup X_2 \ldots \cup X_K$ that minimizes the cost function: $\emptyset(X_1, X_2, \ldots, X_K)$

$$= \frac{1}{K} \sum_{1 \le i \le K} \frac{\sigma_i^2}{\sigma_i^2 + \sigma_0^2} + \frac{K^2}{K^2 + 1} + \frac{1}{K^2} \sum_{1 \le i, j \le K} \frac{\sigma_i \sigma_j}{\sigma_i \sigma_j + \sigma_{ij}^2} \lambda_{ij}$$

The first term seeks *homogeneous* regions; the second seeks minimum K possible. Also, it is useful to consider:

$$\lambda_{ij} = \left| \frac{n_i - n_j}{n_i + n_j} \right|$$

3.2.2 Representation of Segments

The simplest representation of a segmented image uses an array of the form:

$$s(x, y) = \begin{cases} k & \text{if } (x, y) \in \text{Segment } k; \quad \text{where, } 1 \le k \le K \\ 0 & \text{otherwise} \end{cases} \tag{3.1}$$

The value $k = s(x,y)$ is called the label of the segment consisting of the pixel (x,y). The array s is described over the rectangular grid, G, over which the original image is defined, and each pixel is assumed to be a square of finite size. If there are K segments in the image, then the above representation will require an $n^2 \times 1$ vector of integers, each with $\lceil \log_2 K \rceil$ bits. Each entry in the vector corresponds to a specific pixel in the image, and identifies the segment that it belongs to.

In general, the logical interpretation of a segment does not depend on the exact identification number, k, assigned to it. Interchanging the labels of two segments does not change the underlying semantic interpretation. Thus, there are up to $K!$ distinctly different labelings of a single instance of segmentation. In other words, the above representation is not unique, hence not desirable.

Given the fact that we are dealing with an *NP* problem, if we were to solve it by a exhaustive search, it would become impossible to devise a traversal strategy that would remember the solutions that have already been evaluated. The size of the space increases exponentially with the size of the image $n \times n$. A direct representation of each candidate solution takes $n^2 \log K$ bits. Then, it becomes necessary to characterize the population of solutions based on their intrinsic features — in this case substrings of each candidate vector whose length is $n^2 \log_2 K$ bits.

One way to handle large solution spaces is to conduct a randomized search. Genetic algorithm is one such approach. The approach requires that the basis vector used for representing any solution be a Boolean string. Given a potential solution that is currently being verified, the next candidate may be generated by one of the following operations: switching, mutation, or crossover. The Boolean basis must be chosen such that a set of unique physical conditions in the image plane is uniquely in this space as well. Also, each point in this space must correspond to unique segmentation. To facilitate such a method, we chose a Boolean version of an edge map that requires a total of $2n^2 \times 1$ bits. Two bits (Booleans) called e_x and e_y are associated to each pixel and are defined as follows:

$$e_x(x,y) = \big((x,y) \in X_i\big) \wedge \big((x+1,y) \in X_j\big) \wedge (i \neq j)$$

$$e_y(x,y) = \big((x,y) \in X_i\big) \wedge \big((x,y+1) \in X_j\big) \wedge (i \neq j)$$

It takes $O(n^2)$ time to convert between the label-based representation and the Boolean string–based representation. We represent a particular segmentation by its corresponding Boolean string, $\boldsymbol{e}$ and/or the label function $l(x,y)$.

3.2.3 The Basic Structure of Genetic Algorithms

Genetic algorithm is a randomized search approach to solving large-scale optimization problems for which the solution space may be represented by a Boolean string. Let $\boldsymbol{b} = (b_0, b_1, \ldots, b_{2n^2}) \in \mathbb{B}^{2n^2}$ represent an arbitrary segmentation S. Then, at any given instance t, the algorithm maintains a set of candidate segmentations called the *population* at *time t*. The population contains a finite number of solutions, say, $B^{(t)} - \{b_1, b_2, \ldots, b_M\}$. As the computation proceeds in time, this population is said to evolve by a process of natural selection generating a new population at time $t + 1$. The process of natural selection refers to the concept that newly generated solutions are always generated from the best members of the previous generation. The underlying idea is that only the almost optimal solutions will survive after the computation has gone through a large number of iterations, namely, the evolution.

The evolution process is responsible for changing few or all the members of the population at time t into those at time $t + 1$. Three methods may be used in forming a new member. Crossover, mutation, and natural selection are considered in general. To make a contribution to the population a particular member $b_i^{(t)}$ must exhibit large relative fitness, defined as follows:

$$\Pr\left(\boldsymbol{b}_i^{(t)} \in B^{(t+1)}\right) = F\left(\boldsymbol{b}_i;t\right)$$

where

$$F\left(\boldsymbol{b}_i^{(t)}\right) = \frac{\psi\left(\boldsymbol{b}_i^{(t)}\right)}{\sum_{1 \le i \le K} \psi\left(\boldsymbol{b}_i^{(t)}\right)}$$

$$\psi\left(\boldsymbol{b}_i\right) = \text{goodness of the segmentation } \boldsymbol{b}_i$$

The mutation operation is defined as follows. The purpose of this operation is to introduce, from time to time, new possibilities and reduce the effect being locked in local minima. The problem is referred to as inbreeding. The operation $M(i,q;\ t)$: $\boldsymbol{b}_i^{(t)} \rightarrow \boldsymbol{b}j^{(t+1)}$ is carried by inverting the bit q. Note, a solution can coexist with a copy of itself (offspring) which is mutated version. The is, $i \neq j$, in the above.

The crossover operation combines two relatively strong solutions to form a new solution. It does so by choosing a constant p, $1 < p < 2n^2$, and swapping the substrings as follows:

$$\boldsymbol{b}_i^{(t)} = \left(x_0, x_1, x_2, \ldots, x_p, x_{p+1}, \ldots, x_{2n^2}\right)$$

$$\boldsymbol{b}_j^{(t)} = \left(y_0, y_1, y_2, \ldots, y_p, y_{p+1}, \ldots, y_{2n^2}\right)$$

$$\boldsymbol{b}_i^{(t+1)} = \left(x_0, x_1, x_2, \ldots, x_p, y_{p+1}, \ldots, y_{2n^2}\right)$$

$$\boldsymbol{b}_j^{(t+1)} = \left(y_0, y_1, y_2, \ldots, y_p, x_{p+1}, \ldots, x_{2n^2}\right)$$

The method, however, does not work very well on images. A modified crossover operator may be defined by using a parameter vector called $\boldsymbol{p}$. Note, $\boldsymbol{p}$ is also a $2n^2 \times 1$ vector.

$$b_i^{(t)} = \left(x_0, x_1, x_2, \ldots, x_{2n^2}\right)$$

$$b_j^{(t)} = \left(y_0, y_1, y_2, \ldots, y_{2n^2}\right)$$

$$b_i(t+1) = \left(\ldots, \left(p_k x_k \vee = p_k y_k\right), \ldots\right)$$

$$b_i(t+1) = \left(\ldots, \left(\neg p_k x_k \vee p_k y_k\right), \ldots\right)$$

where $\boldsymbol{p}$ is a Boolean vector: $\boldsymbol{p} = (p_0, p_1, p_2, \ldots, p_{2n2})$.

The rational for choosing $\boldsymbol{p}$ is that, in the solution space $\boldsymbol{B}$, edges are correlated. The sequence of edge strokes that constitute a strong line or contour in one solution must wholly be used in another solution. In order to identify such sequences, it is required to form a region adjacency graph (RAG) and a boundary interaction graph (BIG). At this point, two segmentations can also be checked for similarity using graph/subgraph isomorphism tests.

3.3 CURRENTLY KNOWN MULTISTAGE SEGMENTATION PARADIGMS

As described earlier, several heuristics and various hierarchical models have been used for segmentation. In principle, all these algorithms are easily classified into five main categories as (1) bottom-up procedures, based on edges, (2) top-down procedures, (3) semantic-knowledge-based systems; (4) stochastic optimization (randomized search) procedures; and (5) segment inspect-and-revise techniques. Since the control structure of each one is different from the other, each one of them will be explained, to give the idea of how to combine these ideas effectively. The SIRA prototype described here provides a specific representation that facilitates the accommodation of various insights that may be gained from these (algorithmically) very different models of existing segmentation algorithms.

3.3.1 A Bottom-Up Approach Using Edge Detection Techniques

Our implementation of the segmentation is based on an edge detection and boundary formation approach. The process involves many intermediate steps which are described in sequence as follows:

1. *Preprocessing*: The preprocessing does edge-preserving smoothing that reduces random noise and fine textures present in the given image.
2. *Edge Detection and Edge Representation*: This stage basically involves template matching and determining the existence and the strength of an edge at a given pixel.

a. Also addressed in this stage are flexible, spatial representations of an edge under discrete-grid geometry.
b. A simple set of edge evaluation masks is developed.
c. Nonmaxima suppression: as discussed in the foregoing sections, this stage addresses the multiple-edge problem.
d. Relaxation: an iterative procedure to produce a set of globally consistent edges, by increasing or decreasing the strength of each edge element repeatedly until an acceptable global consistency is obtained.

3. *Component Labeling*: Bottom-up procedures tend to produce local results, which holds a promise of global property such as connectedness. The simplest possible approach is to threshold the edge value and apply a labeling algorithm to produce connected regions. If the desired segments are lines, i.e., not closed regions, then one may have to seek contour tracing and/or Hough transforms to consolidate edge points.

3.3.2 A Top-Down Approach Using Split-and-Merge Procedures

The quad tree–based segmentation algorithm consists of the following major groups of operations.

1. Create a quad tree to a depth less than the maximum possible depth defined by the given picture dimensions. Each leaf node represents a square block in the input image.
2. Tree merge the operations in which the four sibling nodes that are currently in leaf level are merged into one, if the uniformity predicates are satisfied by their common ancestor. This procedure is applied repeatedly until there are no four leaf nodes that can be merged any further.
3. Tree split each terminal node in the current tree into four new nodes, if the current node does not satisfy the uniformity predicate.

Note that both of the above techniques transform the initially complete quad tree into an unbalanced quad tree. The underlying structure, however, still remains as a quad tree, i.e., each nonterminal node has exactly four children nodes.

4. Form the initial regions by labeling each terminal node of the resulting cut set of the initial quad tree. At this stage, each region is a square block in shape; however, blocks of different sizes may exist.
5. Form a RAG, and merge adjacent regions (pair-wise), if their union satisfies the uniformity predicates. Repeat step 5 until there are no two spatially adjacent regions that can be merged.
6. Region melt. Collapse the insignificant holes, and small meaningless regions.
7. Output the result, including a facility to merge any regions interactively.

3.3.3 Stochastic Search Techniques

Stochastic search methods such as simulated annealing and genetic algorithm (especially sustained mutation) are expected to perform better for image segmentation.[9]

These algorithms have been applied on several *NP*-complete problems[5,10] that are very similar to image segmentation. In a broad sense, these methods are similar to the hill-climbing algorithm, except that the generation of next potential solution (point) is not always close to the current solution (point) in the search space. That is, statistically speaking, every point in the solution space has a nonzero probability of being selected for the next iteration, regardless of the current state. However, some of them may be preferred more to the others (similar to steepest direction in hill-climbing), based on the current state of knowledge. Also, the system is capable of adaptively modifying and updating its preferences as it completes more and more iterations.

3.3.4 SIRA: Segment Inspect-and-Refine Algorithms

The control structure of a new type of image segmentation algorithms called SIRA is as follows. It combines the advantages of existing top-down and bottom-up methods based on semantic (object-level) models, as well as the atomic and parametric (pixel-level) models of the imaged objects, respectively. They often produce (1) a semantically acceptable but geometrically inaccurate segmentation or (2) a semantically incorrect, but geometrically acceptable (e.g., atomic feature-wise least-mean-squared error) segmentation due to an inevitable loss of edges at a few critical points. Even when the problems are detected, it is required to perform a whole new segmentation, the result of which will not necessarily be the best. This is true, in particular, since it is almost impossible to tune parameters without any trade-offs.

In this approach, the approximate solution will be produced by any of these paradigms, and iteratively refined by: (1) locating and identifying the strong evidence that concurs with semantic- and/or atomic-level-knowledge of the model; (2) propagating them as dynamic constraints to the other levels of segmentation; (3) dynamically splitting (inserting new edges) and regrouping, or merging (deleting edges) the segments; and (4) repeating these steps until the resulting perception concurs at both the atomic and semantic levels.

The numerical values of several aggregated attributes of each segment are used to generate and choose the refinement strategies. It is important to maintain the intermediate segmentation in a form that would facilitate fast insertion and deletion of edges, fast regrouping, splitting and merging of regions, including the reorganization (relabeling) of pixels, fast computation or updating of the useful global attributes required for future decisions. These conditions pose an interesting problem from algorithm design concerning its complexity analysis. The amortized complexity of the dynamic refinement is linear, $O(n^2)$ for $n \times n$ image. It is very suitable for applications where the object models, or the confidence analysis are partially available at the semantic and pixel (parametric) levels.

3.4 APPLICATION OF GENETIC ALGORITHMS TO IMAGE SEGMENTATION

Genetic algorithms (especially sustained mutation) are expected to perform better for image segmentation.[9] These algorithms have been applied on several *NP*-complete problems[5,10] that are very similar to image segmentation.

Let $\boldsymbol{b} = (b_0, b_1, \ldots, b_{2n^2}) \in \mathrm{IB}^{2n^2}$ represent an arbitrary segmentation S. Then, the objective is to minimize, $\phi(\boldsymbol{b})$. Genetic algorithms perform the search for the optimal value $\hat{\boldsymbol{b}}$ iteratively over time. At each iteration, the algorithm considers a set of candidate solutions called a *Population*, say, $\boldsymbol{B}^{(t)}$, at *time t*. The iterative process is commonly referred to as an *evolution* of the population, described as *evolution*: $\boldsymbol{B}^{(t)} \rightarrow \boldsymbol{B}^{(t+1)}$; $\forall_t$. Only the *almost optimal* solutions survive at the end, and they share some common properties called *schemas*. In general, for any given iteration t, roughly 50% of the population $\boldsymbol{B}^{(t)}$ is inherited into the next generation. Also,

$$\Pr\left(\boldsymbol{b}_i^{(t)} \in B^{(t+1)}\right) - \frac{\phi\left(\boldsymbol{b}_i^{(t)}\right)}{\Sigma_i \phi\left(\boldsymbol{b}_i^{(t)}\right)}$$

where $\phi(\boldsymbol{bi})$ = goodness of the segmentation $\boldsymbol{b}_i$.

One way of computing $\psi(\boldsymbol{b})$ is to find the sets $X_1, X_2, \ldots, X_K$ from $\boldsymbol{b}$ by a labeling algorithm and then computing $\phi(X_1, X_2, \ldots, X_K)$. This takes $O(Kn^2)$ time. However, a similar measure could be defined as

$$\sum_{(x,y)} \left(f(x+1,y) - f(x,y)\right)^2 \left[-e_x(x,y)\right] + \sum_{(x,y)} \left(f(x,y+1) - f(x,y)\right)^2 \left[-e_y(x,y)\right]$$

This measure, when minimized, indicates smooth segmentation over a nontextured segment.

The *mutations* ensure that the evolution does not suffer from heavy *inbreeding* and *avoids* local minima. The probability of generating a new solution $\boldsymbol{b}_i^{(t+1)}$ by mutation is kept minimal, say 0.05. Crossover is another important operation that contributes to the controlled diversity of the population of newly generated solutions. The simplest type of crossover operation, called *single pivot crossover,* may be defined as, $C(i,j,p;\, t)$: $(\boldsymbol{b}_i^{(t)}, \boldsymbol{b}_j^{(t)}) \rightarrow (\boldsymbol{b}_i^{(t+1)}, \boldsymbol{b}_j^{(t+1)})$. Thus, the crossover transforms

$$\boldsymbol{b}_i^{(t)} = \left(x_0, x_1, x_2, \ldots, x_p, x_{p+1}, \ldots, x_{2n^2}\right)$$

$$\boldsymbol{b}_j^{(t)} = \left(y_0, y_1, y_2, \ldots, y_p, y_{p+1}, \ldots, y_{2n^2}\right)$$

into

$$\boldsymbol{b}_i^{(t+1)} = \left(x_0, x_1, x_2, \ldots, x_p, y_{p+1}, \ldots, y_{2n^2}\right)$$

$$\boldsymbol{b}_j^{(t+1)} = \left(y_0, y_1, y_2, \ldots, y_p, x_{p+1}, \ldots, x_{2n^2}\right)$$

The algorithm eventually stops, after a large (predetermined) number of iterations, or if convergence has been identified. A population is generally considered converged, if almost all the members of that generation are equally good.

3.4.1 Basic Control Structure

Practical implementation of genetic algorithms involve the following stages: (1) identify a suitable coding strategy that permits binary-string representation of any point in the search space; (2) define a fitness evaluation function to measure the relative merits of any candidate solution over the others; (3) design a suitable data structure to maintain a set of candidate solutions, called population, at any given time; (4) well define operators that will modify some members of the current population resulting in a new generation (future population) using a genetic prototype such as mutation, inversion, and crossover, etc.; and (5) evaluate the generation as a whole and decide the termination of the evolutionary search for the optimal solution. The best solution of the last generation is considered the optimal solution.

The mutation operator requires that very little (minor, local) changes made in the solution strings be able to cause an improvement in the segmentation. Depending on the context, mutation can be performed by (1) inserting a 1 in some arbitrary place in the solution string, by replacing a 0 in that position, or (2) the reverse. The convergence may be accelerated by choosing the positions (pixels) where it is likely to improve the solution. However, every bit position in the string must have a nonzero probability of being chosen for mutation.

Under certain conditions, when a mutation occurs at two places, it effectively results in a net change called a *switch* operation. Switch operations cause no serious changes in the semantic interpretation of two consecutive images. However, they provide a mechanism to improve the exact positioning of the boundary pixels. The method described here assigns a nonzero probability of choice to all pixels for mutation, while it also takes into account a bias that is developed to assess the likelihood of a pixel being an edge pixel.

Any previous knowledge, if available, manifests into joint (conditional) probability among: (1) two cooperating and spatially adjacent or (2) evidence-based noncontinuous groups of pixels being chosen for mutation operations. For example, if a particular bit is inverted from 1 state to 0 state, then all the other bits that are strongly correlated with it are under a threat of becoming useless in the next iteration. The situation, in one sense, must be permitted in order to support the underlying nature of stochastic searches. The spatial interdependence of these bit positions makes any conventional crossover operator almost useless.

We describe a new way of modeling the crossover operators using gate functions. These gate functions are to be chosen from a huge space of up to $O(2^{2n^2})$ binary strings. The choice among various strings can be tuned to be selective based on semantic knowledge of the scene.

3.4.2 String-Based Representation of the Search Space

The input image is described over a rectangular grid, G, of the form: $G = \{x,y) | 0 \leq x,y < n\}$. Each pixel, say, $(x,y) \in G$, is considered a square area of finite size. If there are K segments in the image, then the direct representation of its segmented version would require an $n^2 \times 1$ vector of integers, each with $\lceil \log_2 K \rceil$ bits. Each entry in the vector corresponds to a specific pixel in the image and identifies the segment that it belongs to. The interpretation of the segmented image does not depend on the exact identification number assigned to each segment. Interchanging the labels of two segments does not change the underlying semantic interpretation. Thus, there are up to $K!$ distinctly different labelings of a single instance of segmentation. A different scheme is desired.

Consider a pixel in the grid of image pixels. It has exactly four pixels sharing its boundary. If each pixel is considered a vertex of a graph, then the entire image can be described by a planar graph where each node can have at most four neighbors. Then, each segment in the segmented image can be represented by a connected component of vertices defined as follows. A given partitioning of G is easily represented by two Boolean variables ex and ey as

$$ex[i,j] \leftarrow s(i,j) \neq s(i+1,j) \quad \text{and} \quad ey[i,j] \leftarrow s(i,j) \neq s(i,j+1) \tag{3.2}$$

Both ex and ey play the role of *edges* between two vertices in the graph, which assume only a binary cost function. Also, comparison of two instances of segmentation of the given image can be easily realized in $O(n^2)$ time. Let S_1 and S_2 be the two different segmentations to be compared. Then, these are considered similar if and only if:

$$\sum_{i=1}^{i=n}\sum_{j=1}^{j=n}\left[\left(ex_1[i,j] \oplus ex_2[i,j]\right)+\left(ey_1[i,j] \oplus ey_2[i,j]\right)\right]=0 \tag{3.3}$$

where

$$ex_1[i,j] \leftarrow S_1(i,j) \neq S_1(i+1,j), \quad ey_1[i,j] \leftarrow S_1(i,j) \neq S_1(i,j+1)$$

$$ex_2[i,j] \leftarrow S_2(i,j) \neq S_2(i+1,j), \quad ey_2[i,j] \leftarrow S_2(i,j) \neq S_2(i,j+1)$$

and $\oplus$ means a one-bit exclusive or operator.

Given both ex and ey, the underlying segmentation $s(x,y)$ can be found using standard labeling algorithms, within $O(n^2)$ time. The functions ex and ey represent if a particular edge element actively participates in a boundary between two segments or not.

Consider two Boolean variables, defined as follows:

$$bvx[i,j] \quad \rightsquigarrow \quad s(i,j) \neq s(i+1,j)$$

$$bvy[i,j] \quad \rightsquigarrow \quad s(i,j) \neq s(i,j+1)$$

Both bvx[..] and bvy[..] could be easily obtained by a simple edge-detect operation, followed by thresholding of the edge strength. Let $s(x,y)$ be developed by the labeling algorithm based on $bvx[i,j]$ and $bvy[i,j]$. Then, it is necessary that $bvx[i,j] = true$; to assert that $s(i,j) \neq s(i+1,j)$, however, the knowledge that "$bvx[i,j] = true$" is not sufficient to assert the same.

Even when $bvx[i,j]$ has been set to 1 state in the binary string bvx, the pixels (i,j) and $(i+1,j)$ may inevitably receive the same labels. The inevitability will arise if there is a lack of global support (elsewhere in the image) to give rise to a closed contour consisting of $bvx[i,j]$. That is, the absence of a closed contour passing through the edge element $bvx[i,j]$, the element $bvx[i,j]$ is of no influence on the final partitioning of the image. Any iterative refinement strategy must be aware of this and be prudent if refinements must be activated by suitably changing bvx or bvy.

Let,

$$\widehat{E_x} = \{(i,j) \mid ex[i,j] = true\}$$

and

$$\widehat{B_x} = \{(i,j) \mid bvx[i,j] = true\}$$

and E_y and B_y be defined likewise. Then,

$$\widehat{E_x} \subset \widehat{B_x} \quad \text{and} \quad \widehat{E_y} \subset \widehat{B_y}$$

3.4.3 Generation of the Initial Population

In principle, stochastic search algorithms are insensitive to the initial choice of candidate solution. They are expected to consider every possible solution with a nonzero probability as a candidate for the next iteration. Consequently, a bad strategy in choosing the initial population is not likely to eliminate the ability to find the true solution, except that the convergence may be significantly delayed. However, the initial population must satisfy sufficient diversity among its members, in order to make the search more effective. Also, knowledge available from other sources may be used to accelerate the convergence, by choosing the initial population somewhat statistically biased toward the exemplar that concurs with external knowledge or the believed solutions.

It is reasonable to assume that, in a given string $\boldsymbol{b}$, the bits that correspond to pixels indicating stronger contrast between adjacent pixels are more likely to be turned on. Given, say, 100 strings in a population, then the higher the contrast at pixel (i,j) implies, that the bits $bvx[i,j]$ and/or $bvy[i,j]$ would be turned on among that 100 candidates more frequently. Also, in the case of mutation, it may be useful to compare the relative preference of one bit position over the other. This may be achieved by taking into account the *spatial distribution* of the *local contrast* between adjacent pixels. That is, a mutation operation of $0 \rightarrow 1$–type transitions must frequently be biased to those pixel positions where the contrast is higher. Likewise, the mutations of $1 \rightarrow 0$–type transitions must be biased more toward pixels where there is more homogeneity among local neighbors.

Thus, it becomes important to use a "random event" generator, which, if necessary, can also be biased to some events more than others. A number of random sequence generators are available in the literature. We warn readers against using the standard function *rand* available in garden-variety *C* compilers and Unix systems. These functions are limited to generating a *uniformly distributed* continuous random variable $\alpha(t)$, in the semiopen interval $\alpha \in (0.0,1.0]$. However, it is desired that the random variable be generated in a somewhat biased manner. Let $x_1, x_2, \ldots, x_N$ be N discrete events, whose discrete probability, or plausibility, is given by the numbers, $p_1, p_2, \ldots, p_N$. First, we define a new function $P[i]$, $1 \le i \le N$, such that $P[i] = \Sigma_{j=1}^{i} p_j$, commonly referred to as the cumulative distribution function of the discrete random variable x. It is required to define $P[0] = 0.0$ and $P[N + 1] = 1.0$; then it is possible to map a particular instance, say $\hat{\alpha}$, of the continuous random variable α to a specific outcome $\hat{x}_i$ in the discrete space of $X = x_1, x_2, \ldots, x_N$ using the identity:

$$\hat{\alpha} \in (0,1] \rightarrow \hat{x} = x_i \in \left[x_1, x_2, \ldots, x_N\right]; \quad \text{where} \quad P[i-1] \le \hat{\alpha} < P[i]$$

This mapping is realized using a binary search within $O(\log_2 N)$ time. Also, the bias can be represented by a linear array, whose storage complexity is $O(N)$. The method is useful when the relative preference of all the discrete outcomes, say, $x_i \in [x_1, x_2, \ldots, x_N]$, remain constant in time. However, dynamically updatable bias-tables described in Reference 11 can be used. The method requires explicit binary trees and gives an amortized complexity of $O(\log_2 N)$ for both update and map operations.

First, a simple gradient operator is applied to the image, thus creating two array $p_x[i,j]$ and $p_y[i,j]$ corresponding to the Boolean base variables *bvx* and *bvy*. The numerical values of these coefficients may be expressed by

$$p_x[i,j] = \min\left(\sigma_c, \left|\nabla_x f(i,j)\right|\right)$$

$$p_y[i,j] = \min\left(\sigma_c, \left|\nabla_y f(i,j)\right|\right)$$

where σ_c represents the uncertainties due to camera noise. The noise in general is modeled by a zero mean, Gaussian white noise of variance σ_c^2. It also makes sure that every pixel gets a nonzero preference of being selected as a candidate position in a binary string ***b***, while it is being created and/or modified. Then, the initial population of, say, P strings may be generated using a simple scheme as follows: (1) use the biased random number generator, with $p_x[i,j] + p_y[i+j]$ as the bias, and select a number of points (at least $0.4n^2$), $(\hat{i},\hat{j})$ for each string; (2) at each selected $(\hat{i},\hat{j})$, if desired, draw another random variable, which can be used to decide if $bvx[i,j]$ and/or $bvy[i,j]$ must be set to 1 or 0. It is expected that, in typical applications, the number of boundary pixels is about 20% of the total number of pixels, n^2, in the image. If larger values of p_x and p_y values at a few isolated points tend to dominate, then these values may be clipped at a suitably chosen approximation of the acceptably maximum preference.[12]

3.4.4 Possible Choice for Fitness Functions

Fundamentally, there are two ways to define the fitness measure of an arbitrary instance of segmentation. It is assumed that the instance to be considered is defined by *bvx* and *bvy* from which *s(x,y)* has been already identified, using an $O(n^2)$ labeling algorithm. This algorithm also outputs the segments $S_1, S_2, \ldots, S_K$, each as a list of pixels. Then, an evaluation function to indicate a good segmentation may be defined as

$$\phi(\boldsymbol{b}) = \frac{1}{K}\sum_{1\le i\le K}\frac{\sigma_i^2}{\sigma_i^2+\sigma_0^2} + \frac{K^2}{K^2+1} + \frac{1}{K^2}\sum_{1\le i,j\le K}\frac{\sigma_i\sigma_j}{\sigma_i\sigma_j+\sigma_{ij}^2}\lambda_{ij}$$

The first term seeks *homogeneous* regions and the second seeks minimal values of K. Also, λ_{ij} forces a negative preference with regards to too many small regions adjacent to each other, when defined as follows,

$$\lambda_{ij} = \left|\frac{n_i - n_j}{n_i + n_j}\right|$$

may be useful.

It is assumed that $\sigma_0^2 = \sigma_c^2$. Also, σ_i^2 as well as σ_{ij}^2 are pixel intensity variances computed over the sets, S_i and $S_i \cup S_j$, respectively. The function defined above does not take into account the edges that have been lost in the labeling process due to a lack of support. Consequently, strings that may eventually lead to a better segmentation over the next few generations go undetected and hence are not encouraged to endure. This is due to the many-to-one nature of $\boldsymbol{b} \rightarrow s$ mapping. One way of being selective is to consider another function:

$$\sum_{i=1}^{n}\sum_{j=1}^{n} c_{00}(x,y)\big[(1-bvx[i,j])p_x[i,j] + (1-bvy[i,j])p_y[i,j]\big]$$

$$+\, c_{10}(x,y)\Big[bvx[i,j](p_{\max} - p_x[i,j]) + bvy[i,j](p_{\max} - p_y[i,j])\Big]$$

$$+\, c_{11}(x,y)bvx[i,j]\,\frac{\sigma_l\sigma_m}{\sigma_l\sigma_m+\sigma_{lm}^2} + c_{11}(x,y)bvy[i,j]\,\frac{\sigma_l\sigma_n}{\sigma_l\sigma_n+\sigma_{ln}^2}$$

where $(i,j) \in S_l$, $(i+1,j) \in S_m$, and $(i,j+1) \in S_n$. Also,

$$c_{00}(x,y) = q_1 c(x,y)(-bvx[x,y] \wedge \neg ex(x,y))$$

$$c_{10}(x,y) = q_2[1-c(x,y)]\big((bvx[x,y] \vee bvy[x,y]) \wedge \neg ex(x,y)\big)$$

$$c_{11}(x,y) = q_3[1-c(x,y)]\big((bvx[x,y] \wedge ex(x,y)) \vee (bvy[x,y] \wedge ey(x,y))\big)$$

The function $c(x,y)$ is initially set to 1.0, $\forall(x,y)$. The values q_1, q_2, and q_3 may be adjusted to make the system more tuned to (1) places we would like to have an edge; (2) allowability of spurs and broken edges; (3) places where we want smoothness, respectively. For example, if it is known that the image has been grossly underseg-mented, then the value q_2 must be allowed to be very small; to encourage the evolution, try various possible mutations concurrently. If an *a priori* knowledge of the expected segmentation is available, then $c(x,y) = 1$ refers to those pixels, whose neighborhoods are expected to have a boundary passing through them. In particular, for motion analysis systems involving image sequences of slowly varying dynamic scenes, $c(x,y)$ can be derived by a convolving a spatial (uncertainty) mask with a binary-valued image constructed using $(ex[i,j] \vee ev[i,j])$ of the best segmentation of the previous scene. In the case of evolutionary knowledge about the past few generations may be used to update $c(x,y)$. Furthermore, schemas that are likely to be present generally correspond to the string arcs (consisting of several edge elements) in the boundary interaction RAGs. This can also be computed and updated effectively.

3.4.5 The Switch Operator

Consider an operation that involves removing a pixel $(i,j) \in S_l$ from S_l and assigning it to another region, S_m. It is assumed that the segments S_l and S_m are physically adjacent, and hence connected to each other by an arc in the BIGRAG graph. Then, incrementally refining the boundary involves such an operation. Essentially, this amounts to at least two $0 \rightarrow 1$–typed, and two $1 \rightarrow 0$–typed mutations. There will be at most three pairs of such $1 \rightarrow 0$ and $0 \rightarrow 1$ types of mutations. These are more useful than just a single-bit mutation. The single-bit mutations are required when the segmented image shows fewer segments than expected. Especially when the semantic structure of the segmented image is acceptable, then mutations of switch types prove to be more useful.

3.4.6 The Crossover Operators by Gate Functions

The conventional definition of crossover was already discussed in this chapter. Let $\boldsymbol{g}$ be a gate function, such that

$$\boldsymbol{g} = \left(0,0,0,\ldots,g_p = 1,1,\ldots 1\right) \quad \text{where} \quad g_i = \begin{cases} 0 & \text{for } 1 \le i < p \\ 1 & \text{for } p \le i \le 2n^2 \end{cases}$$

Then, the crossover can be compactly described by:

$$\boldsymbol{b}_i^{t+1} \leftarrow \left(\neg\, \boldsymbol{g} \wedge \boldsymbol{b}_i^t\right) \vee \left(\boldsymbol{g} \wedge \boldsymbol{b}_j^t\right)$$

$$\boldsymbol{b}_j^{t+1} \leftarrow \left(\boldsymbol{g} \wedge \boldsymbol{b}_i^t\right) \vee \left(\neg\, \boldsymbol{g} \wedge \boldsymbol{b}_j^t\right)$$

Choosing p as the pivot parameter, there are exactly $2n^2$ gate functions that may be developed. However, a more-modified approach may be to choose any number of contours, or line segments, in the image planes in the form of gate functions. For instance, a list of all the edge elements that participate between any two (only two) regions in the BIGRAG of each segmentation may be considered a gate function. Or a streak of line that may be extracted from the image by a simple thresholding type of segmentation can be used, as well. Also, using morphology operators, dilate, and expand, one can generate a set of equivalent gate functions, to promote inter-region boundary adjustments.

3.5 EXPERIMENTAL RESULTS

The methods described in this chapter were implemented using an X-window-based image-processing environment. The images were digitized from a Pulnix camera, on an Imaging Technology ITI-150 series system. The image was originally resampled to compensate for aspect ratio of 4:5. Also, to enhance the speed, each image was cropped to 128 × 128 pixels. The camera variance due to digitizer noise was considered 0.5, and the lighting was considered uniform over the entire objects.

Based on our knowledge of the objects, we anticipate about ten long lines, and four segments in the case of the cube. We also anticipate that each line be roughly 64 pixels, about half of image dimensions. This is well within the reach of $0.4n^2$, i.e., 4096 bits to be randomly chosen among the initial population of 128. Note, we do not maintain the segmentation of each one of the 128 strings. However, only one solution is labeled at any given time. Each time a candidate is labeled and evaluated, the string as well as its goodness values are stored in a priority-queuelike structure. For larger population sizes, this would have to be accommodated into B^*-like trees, where the strings may reside in disks. A heap is used so as to find the worst solutions at any given time — to be replaced by the newly generated ones.

After generating the first 128 candidate solutions, one can now initiate the genetic algorithmic search. Now, we randomly choose a few of the members in the population for mutations. And a few of the candidates for crossover, etc. To do the crossovers and the sustained mutations, it was necessary to know a family of constrained gate functions. A pool of roughly 64 gate functions was maintained, with a certain preference associated to it. Initially, the list was empty. After each candidate in the initial population was segmented, it was possible to identify the sequence of edge elements that directly participated in the resulting segmentation of that candidate. For each one of them, the preference value was computed based on the length of the sequence and the contrast between the segment it separated. Only the most dominant strings were maintained, and it is not clear if only 64 strings are enough. The input images and their segmentation are illustrated in Figure 3.1 through Figure 3.4.

FIGURE 3.1 Input image.

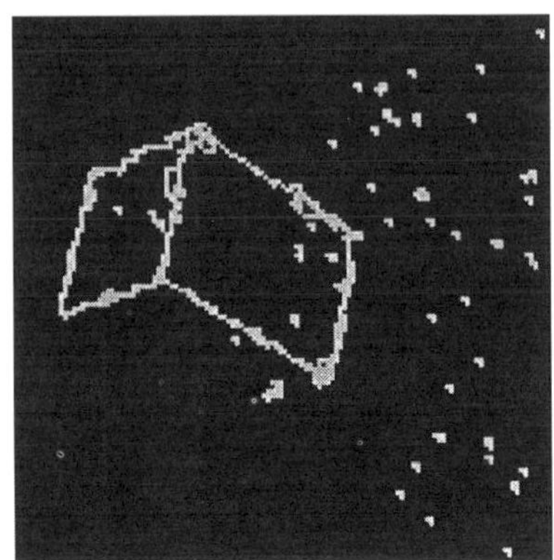

FIGURE 3.2 Segmented image.

FIGURE 3.3 Input image.

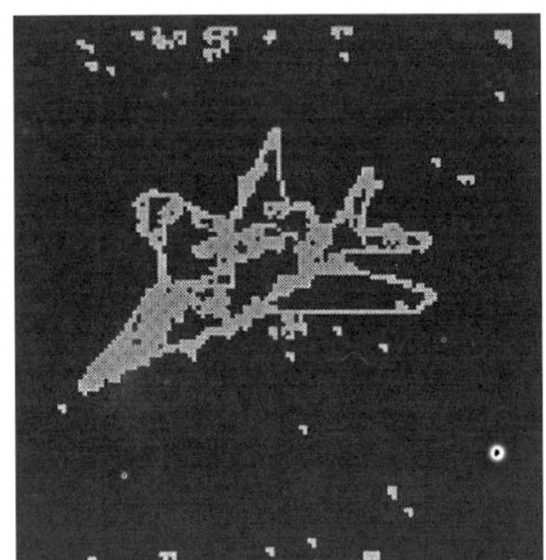

FIGURE 3.4 Segmented image.

REFERENCES

1. Pavlidis, T., *Structural Pattern Recognition,* 2nd ed., Springer-Verlag, New York, 1981.
2. Brice, C. and Fennema, C., Scene analysis using regions, *Artif. Intelligence*, 1(3), 205–226, 1970.
3. Horowitz, S. L. and Pavlidis, T., Picture segmentation by a tree traversal algorithm, *J. Assoc. Comput. Mach.,* 23, 368–388, 1976.
4. Feldman, J. A. and Yakimovsky, Y., Decision theory and AI: a semantic based region analyser, *Artif. Intelligence,* AI-5(4), 349–371, 1974.
5. Talbi, E. G. and Bressiere, P., A parallel genetic algorithm applied to the mapping problem, *SIAM News Lett.,* April 1991.
6. Prager, J. M., Extracting and labeling of segments in natural scenes, *IEEE Trans. Pattern Anal. Mach. Intell.,* 2, 16–27, Jan. 1980.
7. Greogy and Gardner, M., *Seeing: Mind's Eye,* Addison–Wesley, New York, 1980.
8. Young, T. Y. and Seetharaman, G., A region-based approach to tracking 3D motion in an image sequence, *Adv. Comput. Vision Image Proc.,* 3, 1986.

9. Petry, F. E. et al., Scene recognition using genetic algorithms with semantic nets, *Pattern Recognition Lett.*, 11(4), 1990.
10. Jones, D. M., Clustering with Genetic Algorithms. Technical Report GMR-TR-7156, General Motors Research Laboratories, Warren, Michigan, 1989.
11. Seetharaman, G. and Rao, K., A Versatile Random Number Generators for Non-Standard Dynamically Updatable Probability Distribution. Technical Report CACS-TR-93, The Center for Advanced Computer Studies, University of Southwestern Louisiana, Lafayette, 1993.
12. Zucker, S. W., Region growing: childhood and adolescence, *Comput. Graphics, Vision Image Process.*, CVGIP-5(3), 382–399, 1976.

4 Cancer: A Challenge for Control Theory and Computer Modeling

Werner Düchting, Waldemar Ulmer, and Thomas Ginsberg

CONTENTS

Abstract — The goal of this chapter is to provide insights into how system analysis, control theory, and computer science can stimulate new approaches to interpret cancer as a structurally unstable closed-loop control circuit, to simulate temporal and spatial normal cell renewal and tumor growth, and to optimize cancer treatment protocols via computer simulation prior to clinical therapy. Based on control theory and cell kinetic observations, our group has constructed cell cycle models at the cellular level describing the cell growth and regulation mechanisms of normal and malignant cells. The models have been transformed into computer algorithms. By that, cell growth kinetics and the related spatial growth (2D, 3D) could be studied under different conditions and perturbances. The simulation experiments were performed for tumor spheroids, but capillary networks were also considered. Treatment models for surgery, chemotherapy, and radiotherapy were developed in order to compare different treatment methods and schedules with regard to the optimization of treatment protocols. The simulation results are in good agreement with empirical data. Computer simulations offer not only tests of hypotheses or comparisons of different treatment schedules, but also save costs by partial substitution of long and expensive biological test series by simulation experiments (e.g., in pharmacology a reduction of experiments with animals).

0-8493-7962-8/98/$0.00+$.50

4.1 INITIATION AND MOTIVATION

In oncology one characteristic feature of cancer is uncontrolled cell proliferation. This observation has stimulated numerous biomathematicians to construct cell proliferation models (continuous, discrete, deterministic, stochastic) based on, e.g., differential equations describing growth and kinetics of abnormal cell multiplication.[1-11] During the past decade, oncogenes and suppressor and repair genes have shifted the origin of cancer to the molecular biology level, i.e., to genetic alterations.[12-15] By this, a previously proposed, attractive working hypothesis of our group[16] was revitalized to interpret cancer as structurally unstable, negative-feedback control loops. The main question in this scenario is in which way proliferation of normal and abnormal cells is regulated and can be illuminated at different levels: (1) molecular level, (2) cellular level, and (3) organ level. At present, the experimental data about the control mechanisms at the first level are still insufficient. For this reason, we have restricted our considerations on constructing feedback control models which describe the cell division of normal and tumor cells to the cellular level. Before going into details, some general remarks of modeling and simulation shall be given. By modeling, we mean the study of a system using basic physical laws and relationships.[17] Models are only mappings of a more complex reality! So, what we call a system model is actually a reflection of the modeler's understanding of the reality, its components, and their interrelations (Figure 4.1). Therefore, one should keep in mind which aspects have been omitted and what assumptions have been made in modeling. Thus, the computer model is an operational computer program system implementing a system model. The output of a computer run may be the predicted time behavior of the dynamic system. In this sense, simulation can be regarded as the art and science of experimenting with models.[18-21] There are many reasons simulations are valuable: they check and optimize the design of a system before its construction; they help to avoid costly design errors, to ensure safe designs, or to test hypotheses and to vary comfortably the structure and parameter sets of the system. Examples of modern areas of application include space flight simulations, migration in urban systems, as well as pattern formation or spread of epidemics. In contrast to the goals of many biostatisticians our foreground aim is not data-fitting in a simple manner, but stimulation of scientific questions by applying well-established methods of control theory[22] to the regulation mechanisms of the cell division process (Figure 4.2). One crucial question is under which conditions does the negative-feedback closed-loop circuit become unstable.

4.2 MODELING STRATEGY

For studying the process of carcinogenesis, tumors normally are induced to animals or cell cultures, i.e., *in vivo* or *in vitro* models are used. Our approach is concerned with computer models, which means, strictly speaking, with "models of the models."

The computer modeling of highly complex biological systems requires a decomposition of the problem and appropriate simplifications. These are dependent on the ultimate aim of modeling. If one wants to model the spatial surgical removal of a tumor *in vivo,* for instance, in the skin, one has to assume a limited tissue volume,

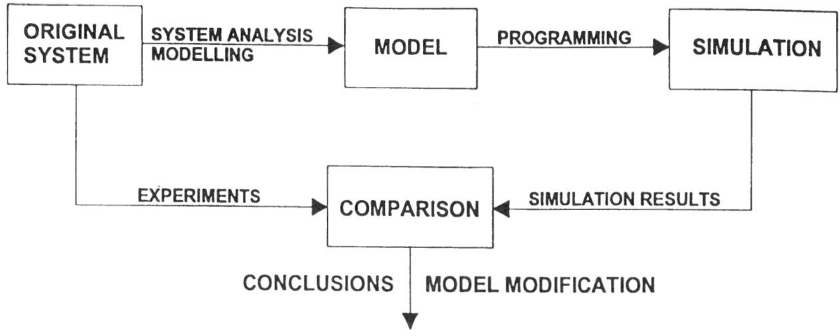

FIGURE 4.1 Steps of modeling.

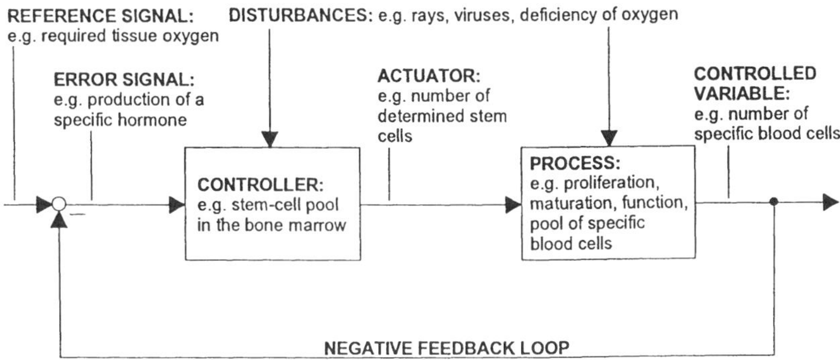

FIGURE 4.2 Schematic illustration of a disturbed closed-loop control circuit describing cell renewal processes.

a constant volume of a cubic cell, a homogeneous tissue structure, and only horizontal and vertical correspondence to neighboring cells mainly caused by restrictions of computer capacity. At the present stage, heterogeneity, immunologic reactions, and the formation of metastases have to be neglected. The computer model is composed by separate models for normal and malignant cells with different cell production rules (e.g., a tumor cell is theoretically capable of unlimited division). These models are connected by a set of cell-to-cell interaction rules which may be position — and energy — dependent. That means if the oxygen or glucose supply of a proliferating cell becomes low, the cell will turn over into the dormant G0 phase. The required experimental data are the initial temporal and spatial configuration of the cell system and the cell cycle times, which are created by a pseudorandom number generator in the computer. For representing the results, a 2D or 3D spatial visualization of normal and malignant tissue is required. If we want to study the temporal and spatial growth of an *in vitro* tumor spheroid (a model of nonvascularized early tumor growth) with the aim of investigating various treatment schedules, additionally we have to construct a cytokinetic model of a tumor cell implementing distinguished cell cycle phase durations (T_{G1}, T_S, T_{G2}, T_M). If we intend to optimize

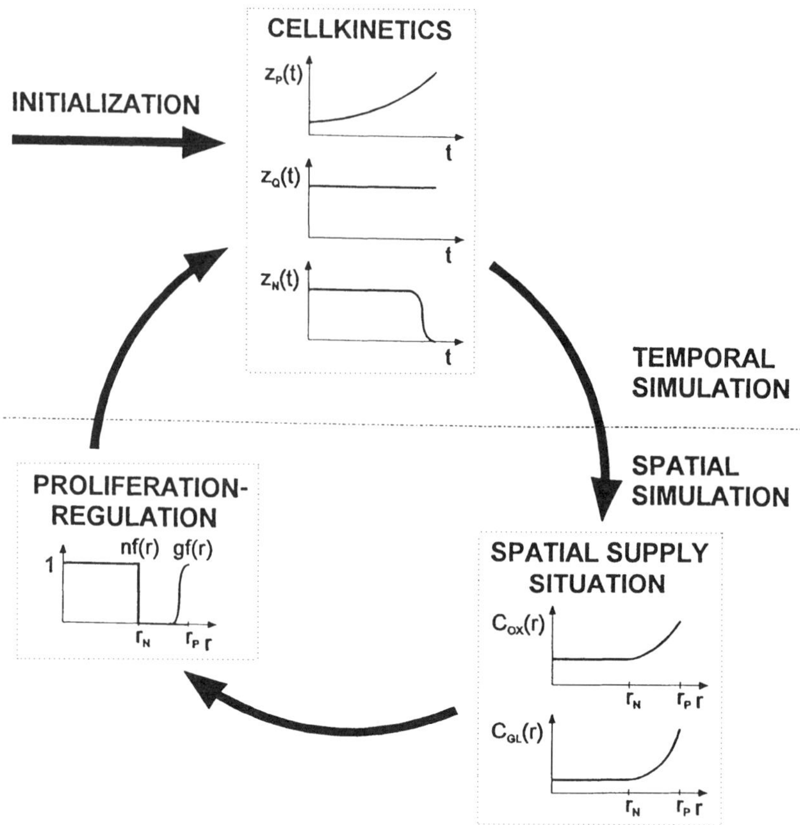

FIGURE 4.3 Growth model of a tumor spheroid. z: number of cells; C: concentration; r: radius of the tumor spheroid; nf(r): necrotic fraction; gf(r): growth fraction; OX: oxygen; GL: glucose; P: proliferating; Q: quiescent; N: necrotic.

cancer treatment, we need a treatment model, and in order to consider side effects on normal tissue we have to model different cell renewal processes of rapidly and slowly proliferating normal tissue. This procedure requires a set of more than 10 variables and about 30 parameters, which may be partially dose dependent. Every scientist knows about the difficulty to get or to estimate these parameters by experiments. Therefore, a crucial bottleneck very often is the lack of sufficient experimental data which are the basis of all modeling activities.

For demonstration, Figure 4.3 represents the growth model of a tumor spheroid.[23-24] In this case, one underlying assumption is that the internal (autocrine) cellular control loop of a cell has become unstable[16,22] and the transformation process from a normal cell to a cancer cell has been finished. It should be mentioned that in Figure 4.3 the time dependence of the proliferation is controlled by the spatial status of oxygen and glucose (described by a system of differential equations related to diffusion processes) introduced by experimental data. Therefore, the simulation run on a VAX workstation with a VAX station 3200 which takes 4 to 6 days is interrupted each 3 h

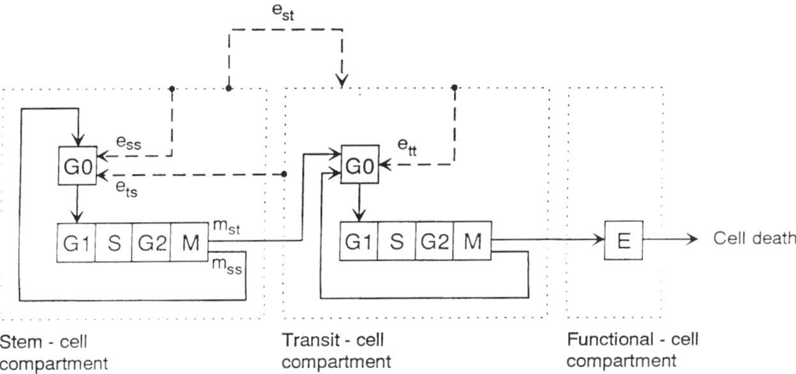

FIGURE 4.4 Cell kinetic model of rapidly proliferating normal tissue. m_{ss}: Share of stem cells which derive from stem cells during mitosis; m_{st}: share of transit cells which derive from stem cells during mitosis; e_{ss}: share of lacking stem cells which stimulates stem cells to leave the dormant phase G0; e_{ts}: share of lacking transit cells which stimulates stem cells to leave the dormant phase G0; e_{tt}: share of lacking transit cells which stimulates transit cells to leave the dormant phase G0; e_{st}: share of lacking stems cells which stimulates transit cells to migrate into the stem cell compartment.

by a simulation loop which checks whether the actual simulation results (cell number in the various stages of the cell cycle) are in accordance with the experimentally gained supply data or not. If not, the proliferation mechanism starts its correction work.

In Figure 4.4 a completely different model is presented.[25] It symbolizes a cell kinetic compartment model describing the time behavior of normal cells after apoptosis, which has disturbed the balance of cells in the distinguished compartments. From the viewpoint of control theory, the modeling of the interacting signals in the communication paths between the different compartments has been performed in the way that the number of cells is tried to be kept constant in the stem cell and transit cell pool[26] via the interacting signals e_{ss}, e_{ts}, e_{tt}, e_{st}, m_{ss}, m_{st} (Figure 4.4).

The complex models described here have been transformed into computer programs — FORTRAN IV or C^{++} and block-oriented simulation languages (Matrix$_x$, MATLAB) have been used. The modeling of different temporal and spatial responses of cell growth to inner and outer perturbances (e.g., chemical agents or radiation dose) allows a better understanding of the various cell regulation mechanisms.

4.3 TUMOR GROWTH

Let us now start with computer experiments simulating the *in vitro* growth of a tumor spheroid.[23] For this purpose, a single tumor cell (in mitotic phase) has been placed in the center of a nutrient medium of a 3D cell space, whose volume amounts to 1 mm^3. The model is based on the cell cycle regulation mechanisms, and the input data are the cell cycle phase durations which are created by a pseudorandom number generator. After growing according to the cell production and interaction rules of the computer model,[27] the spatial 2D steady state is illustrated in Figure 4.5a. A good qualitative agreement with the experiment can be observed in the cross section

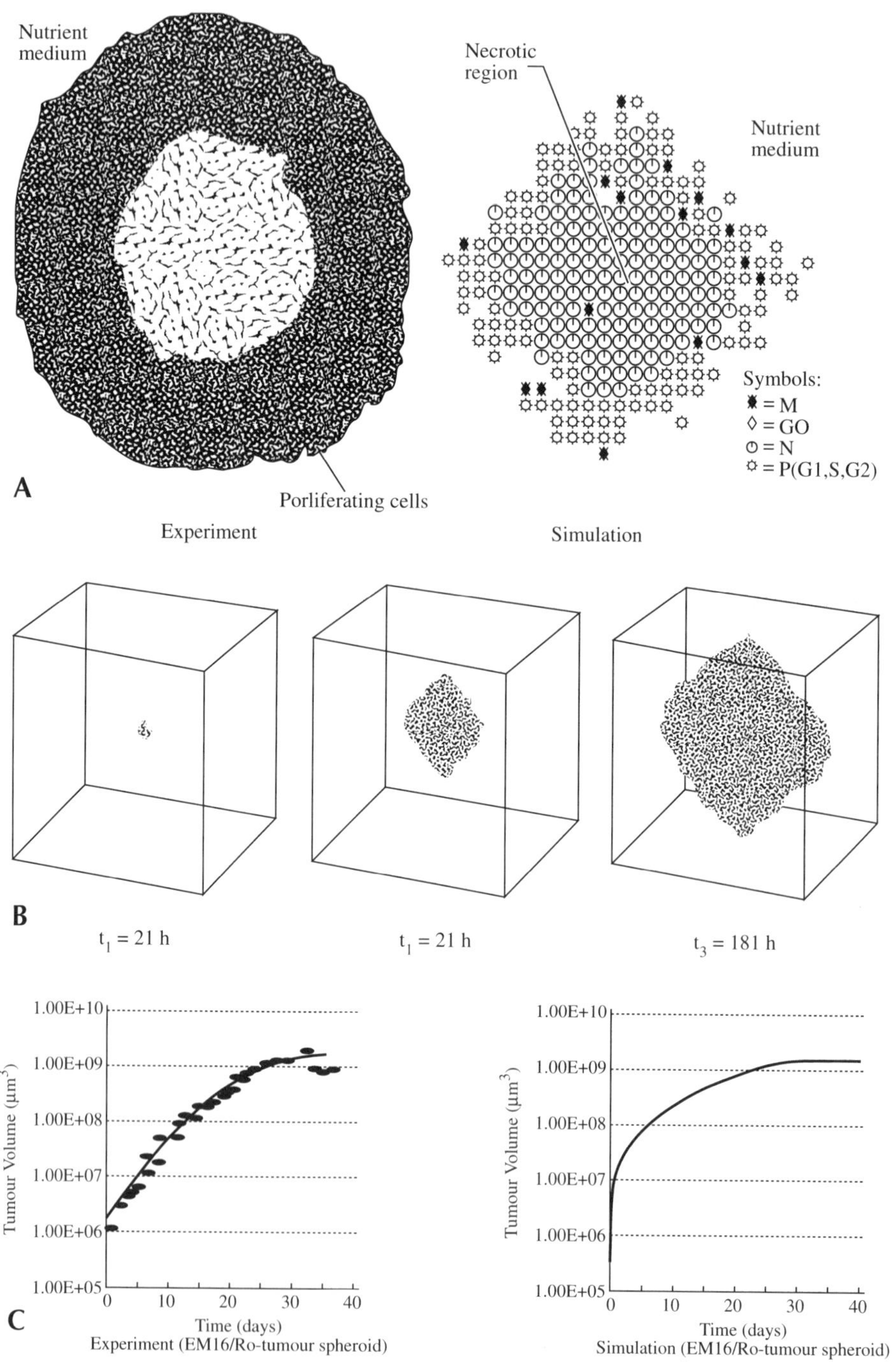

FIGURE 4.5 Tumor growth *in vitro*. (a) 2D cross-section of a tumor spheroid; (b) 3D growth of a tumor spheroid; (c) tumor volume as a function of time.

(Figure 4.5a) as well as via the 3D plot of Figure 4.5b. An improved computer model[24] considering additionally the oxygen and glucose supply in detail leads to a very good quantitative fit of the simulated time course with experimentally gained data of the mamma sarcoma of the mouse (EMT6/Ro-tumor spheroid) given in Figure 4.5c. The transition to the simulation of *in vivo* tumor growth is much more complicated. In a very first step we have neglected the capillary system. Furthermore, in the modeling approach[28] we have assumed that one can only distinguish between normal and malignant cells. Thus, it was possible to study the growth of a tobacco leaf cancer[28] starting from a nucleus of about 20 transformed cells which have been placed in the outside rim of the tissue. The spatial 2D configuration after the tumor has grown up is represented in Figure 4.6a. Parallel to this work we have been thinking about the introduction of the vascular system into our modeling world. In Reference 29 the approach has been made to study the formation of the capillary network of a brain segment of a rat as a regulatory process during the oncogenesis of evolution (Figure 4.6b). Thus, we could simulate in an extended computer model[30] the formation of an *in vivo* micrometastasis starting with a single transformed cell which was inoculated in the direct neighborhood of a capillary (Figure 4.6c). The power of the model may be seen from the 2D plot (Figure 4.6c) in which the possibility of visualizing the neovascularization (tumorangiogenesis effect) is included.

4.4 TUMOR SURGERY

If we want to model cancer treatment we have to extend the models describing tumor growth by developing additional program packages considering the various treatment methods and schedules. A relatively simple case is that of a surgical removal of tumor cells[28] because in our computer model we have to eradicate the undesired cells only. This has been performed in the example of Figure 4.7. In this case, we have restricted the cell space symbolizing a section of the skin to a 2D 100×100 cell matrix consisting of normal cells coded by the symbol 1 which characterizes the specific mean life span of a cell. The growth pattern of a normal cell renewal system is generated by a set of cell production and interaction rules.[28] For some cells, the multistage transformation process of normal into cancer cells has been finished. So the existence of a tumor nucleus according to Figure 4.7a may be assumed.

The multiplication of a tumor cell does not obey production rules for normal cells but a special catalog valid only for tumor cells, which tells that a division of a tumor cell can take place even if there is no vacancy for one or two daughter cells.[28] The input data of our model are the mean life span of normal and tumor cells, the percental cell loss, and the initial configuration of the tumor. The spatial step-by-step increase of the tumor is illustrated in Figure 4.7b. At $T = 50$ units of time most of the tumor cells are removed by a surgical operation. But the remaining tumor cells demonstrate obviously the rapid rise of new tumor cells in Figure 4.7d. Thus, an extremely complex biological system can be simulated near to reality by means of a few simple assumptions. The model enables us to study the influence and variability of the mean life span of a tumor cell, of the size and initial configuration of the tumor nucleus, and of the percental tumor cell loss. One important practical application is the training of surgeons by simulating the operational procedure prior

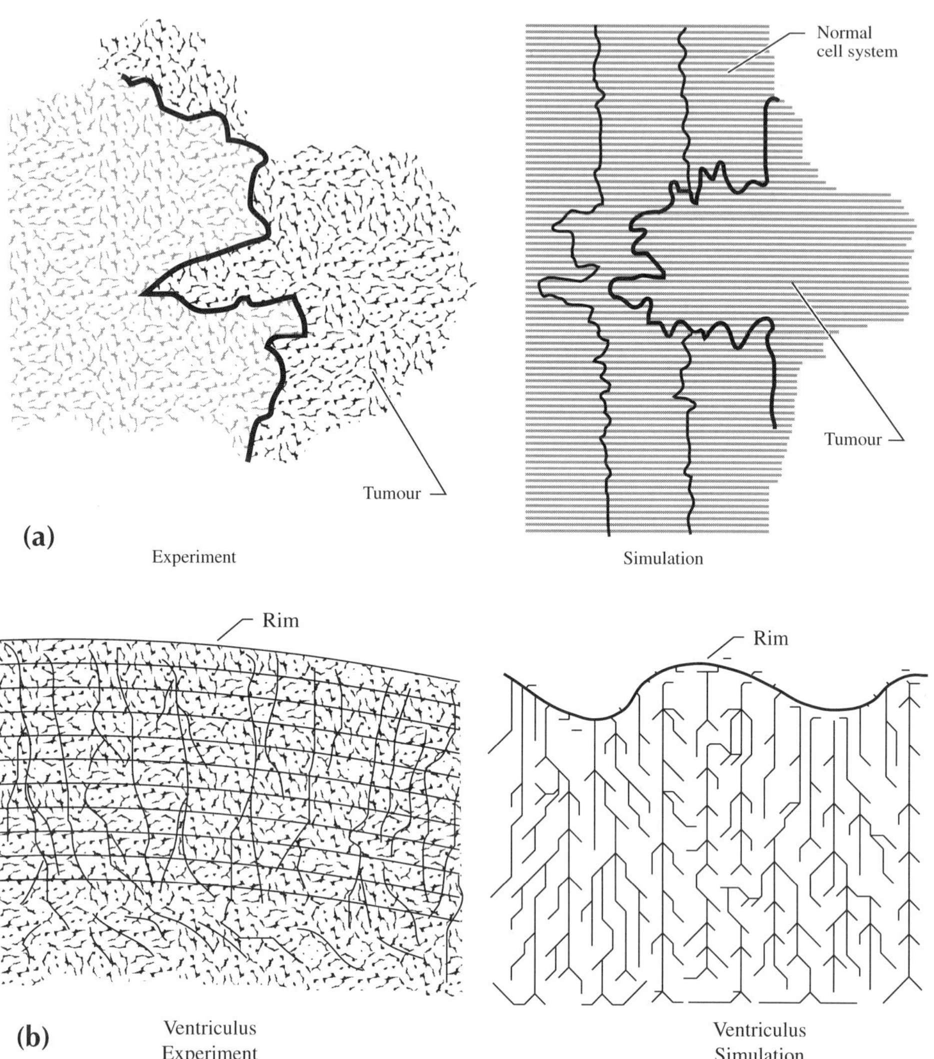

FIGURE 4.6 Tumor growth *in vivo*. (a) 2D growth of a tumor in a tobacco leaf; (b) 2D capillary network of the cortex of a rat; (c) 2D formation of a micrometastasis (simulation).

to clinical treatment; a further one is the importance of the tumor extension considered in the model for microsurgery.

4.5 CHEMOTHERAPY

Cell cultures may not be used to study only the division of tumor cells, but also to determine the cytotoxicity of chemotherapeutic drugs. Many chemotherapeutic agents preferably affect a very particular phase of the cell cycle, which means they act phase specifically.[31,32] Now, tumor cell growth in a cell culture is a very good

FIGURE 4.6(c)

"*in vitro* system" which can be simulated by a computer model.[24,27,33,34] In contrast to the oversimplified model of the previous section, in which one could only distinguish between normal and malignant cells, in this case our model is based on the cell cycle regulation mechanisms that make it possible to differentiate between tumor cells residing in different stages of the cell cycle. Furthermore, cell production and interaction rules allow construction of the actual spatial (2D, 3D) configuration. The input data of the model are cell kinetic data (cell cycle phase durations) and treatment schedules.[27] The effect of drug resistance has not yet been implemented in our model. However, this phenomenon can be taken into account by the dose dependence of the diffusion of a drug through cell membranes (passive permeation) and using methods and results in Reference 24; work in this direction is in progress. The example of Figure 4.8 illustratively demonstrates the application of an agent which has killed all proliferating tumor cells (Figure 4.8b) of the outer rim of the tumor spheroid at $T = 201$ units of time. By this procedure the outer, quiescent G0-tumor cells are recruited into the cell cycle again and tumor growth proceeds according Figure 4.8c. The time at which cytotoxic treatment is given is extremely important. From the time course plotted in Figure 4.8d, one can see the optimal moment when the cytotoxic drug has to be administered for a second time. This is not the time when the total number of tumor cells has reached its minimum (at $T = 250$ units of time in Figure 4.8d), but following our proposal: The drug should be administered at that time when the number of G0-cells has reached its minimum (at $T = 205$ units of time in Figure 4.8d), because then the maximal number of tumor cells resides in the proliferating phase and can optimally be hit by the phase-specific cytotoxic drug. It should be stressed that this approach mainly provides qualitative results which allow testing of new ideas concerning the variation of treatment schedules. In any case, the simulation results have to be confirmed by experiments (*in vitro, in vivo*) and clinical evidence. Based on results of investigations[35-37] on the cytotoxicity of some well-known anticancer drugs monitored in cell cultures (monolayers and spheroids), the present stage of the simulation model[24] is being extended to encompass cancer chemotherapy and the integrated modality of chemotherapy and radiotherapy at the same level as the sole radiotherapy presented in the following section.

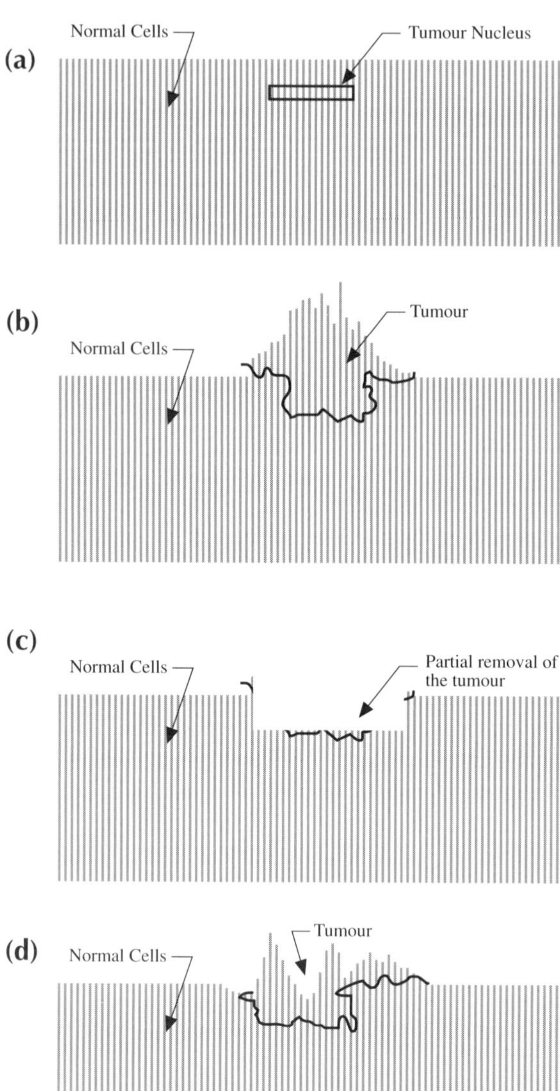

FIGURE 4.7 Surgical removal of tumor cells. (a) Tumor nucleus (initial configuration) at $T = 1$ unit of time; (b) tumor growth (configuration at $T = 25$ units of time; (c) tumor eradication; (d) recidiv configuration at $T = 50$ units of time.

4.6 RADIATION THERAPY

The tumor spheroid model used in this case is based on the cell cycle model described in preceding sections. It has been extended and improved by considering such metabolic factors as oxygen and glucose supply.[24] The simulation of radiation therapy

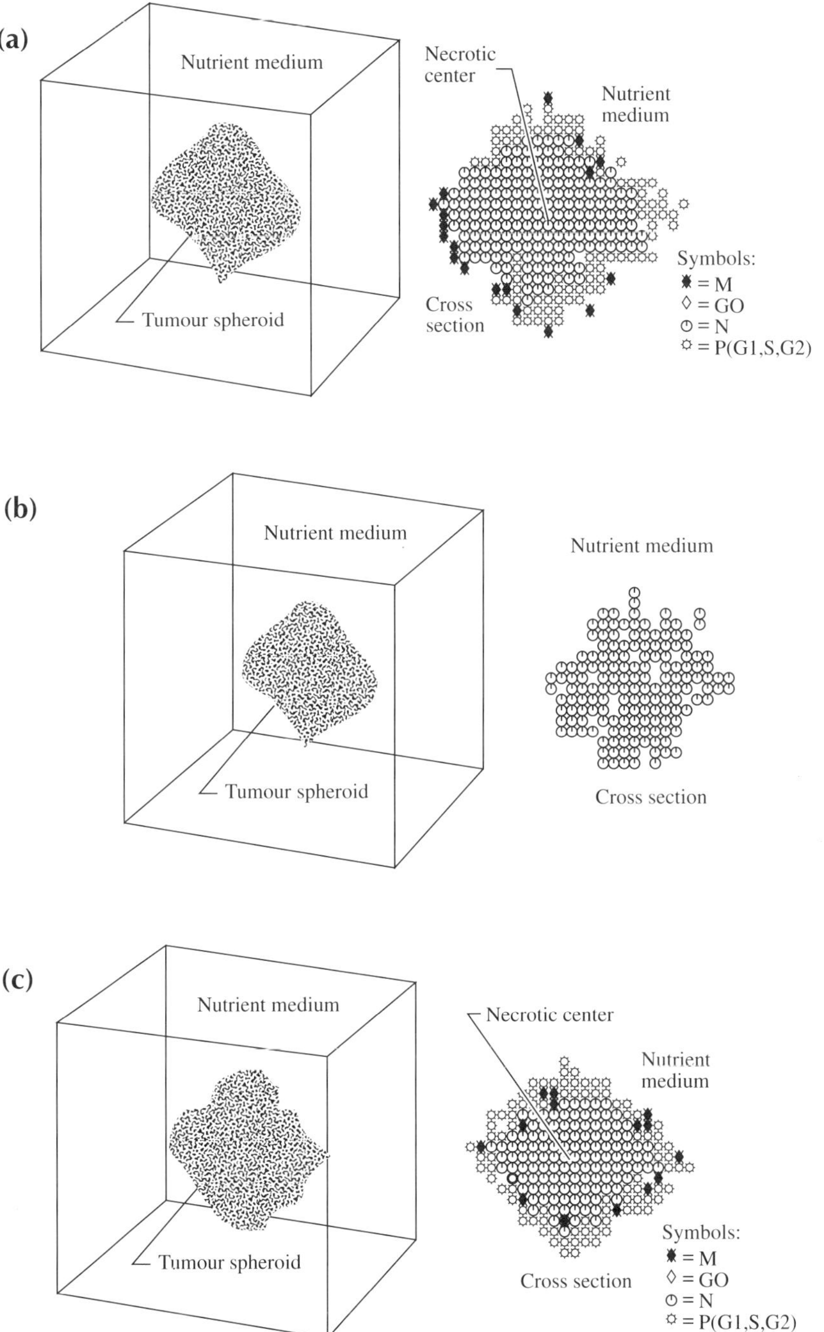

FIGURE 4.8 Chemotherapy applied to a tumor spheroid. (a) Configuration at $T = 200$ units of time; (b) chemotherapy at $T = 201$ units of time; (c) tumor configuration at 100 units of time after chemotherapeutic treatment; (d) number of tumor cells as a function of time.

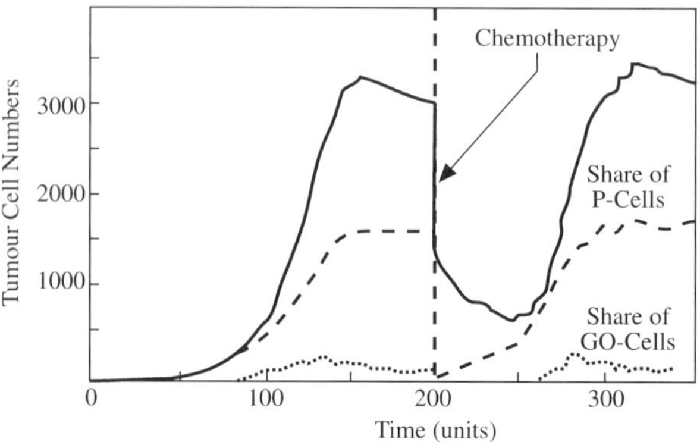

FIGURE 4.8(d)

is much more complicated than that of a surgical removal of a tumor. First of all, an appropriate radiation model is required. In our case we have chosen the linear-quadratic survival function[38,39] as the dose–response relationship of cell cultures (monolayers and spheroids). The aim of radiation treatment is the maximum kill of tumor cells and the minimum damage of normal cells.[40-42] Thus, in a first step we have applied five different clinical treatment schemes[43] (Table 4.1) to the computer tumor spheroid model of the mamma sarcoma of the mouse.[24] The standard fractionation (5×2 Gy per week) has been the usual schedule, and it is still preferred in most hospitals. Currently, there are ongoing clinical studies to evaluate modifications of this standard scheme with regard to an improved tumor cell kill effect and eventually reduced side effects to normal tissue. Therefore, the essential question is what is more favorable, a multifractionated irradiation or an irradiation with a weekly high single dose? In agreement with qualitative clinical observations,[44-46] the following simulation results will show that a general answer cannot be given to this question. Indeed, only a specific solution for each case is possible because, in addition, the side effects of radiation to normal cells have to be considered, too. First, let us look to the irradiation results of the *in vitro* EMT6/Ro-tumor spheroid (of the mamma sarcoma of the mouse) which has been irradiated in Figure 4.9 starting when the tumor has reached its saturation state (see Figure 4.5c) at $T =$ 40 days. Figure 4.9 impressively indicates that only the hypofractionation (Table 4.1) leads to a severe tumor cell kill,[24] but this fact is certainly dependent on the cell line under consideration. Now, the radiation effect to a rapidly regenerating normal cell renewal system has to be studied. If we apply the radiation model to the cell renewal compartment model[25,47] of Figure 4.4, the influence of irradiation to fast-proliferating normal cells can be visualized. As an example of a single-dose irradiation, Figure 4.10 illustrates the response curves of the epidermis of the mouse. The irradiation of the epidermis of the mouse with five different clinical treatment schedules according to Table 4.1 is published elsewhere.[24] But the simulation results tell us that a recovery of the cell system can be expected, unless an accelerated irradiation scheme is applied. A clinical experience is that often the late effects of irradiation are really the dose-limiting factor.[43-46] Therefore, we had to construct a

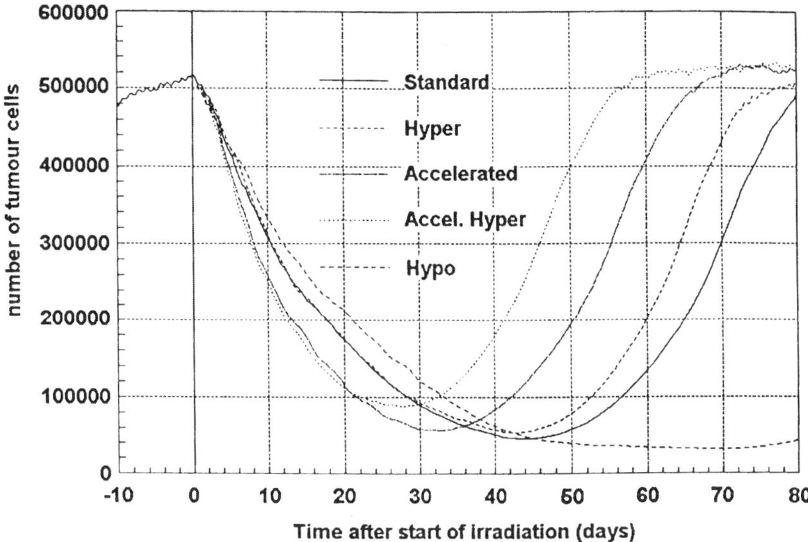

FIGURE 4.9 Irradiation of a EMT6/Ro-tumor spheroid (of the mamma sarcoma of the mouse) with five different fractionation schemes (see Table 4.1).

TABLE 4.1
Fractionation Schemes

Scheme	Dose
Standard fractionation	1 × 2 Gy per day; 5 days per week; 60 Gy total
Hyperfractionation[a]	2 × 1.2 Gy per day; 5 days per week; 72 Gy total
Accelerated fractionation[a]	2 × 2 Gy per day; 5 days per week; 60 Gy total
Accelerated hyperfractionation[a]	3 × 1.5 Gy per day; 5 days per week; 54 Gy total
Hypofractionation	1 × 6 Gy per week; 60 Gy total

[a] Interval between treatments per day: 6 h.

modified computer model,[47,48] which considered the cell kinetic behavior of slowly regenerating normal cell systems (lung, brain) with dominantly quiescent G0-cells. Again, by applying the five different clinical fractionation schemes (Table 4.1) to the lung parenchym of the mouse, Figure 4.11 shows that now the two schedules hyper- and hypofractionation lead to severe (late) damages of the normal cells.[24] In any case, the qualitative simulation results always have to be confirmed by experiments and/or clinical evidence.

Finally, an important advantage of the simulation approach shall be stressed. In the biological reality you can only perform experiments with a tumor embedded in a mixture of rapidly and slowly regenerating normal tissue. By applying computer modeling, it is possible to study the influences of treatment to each tissue separately and to compare the results with each other.

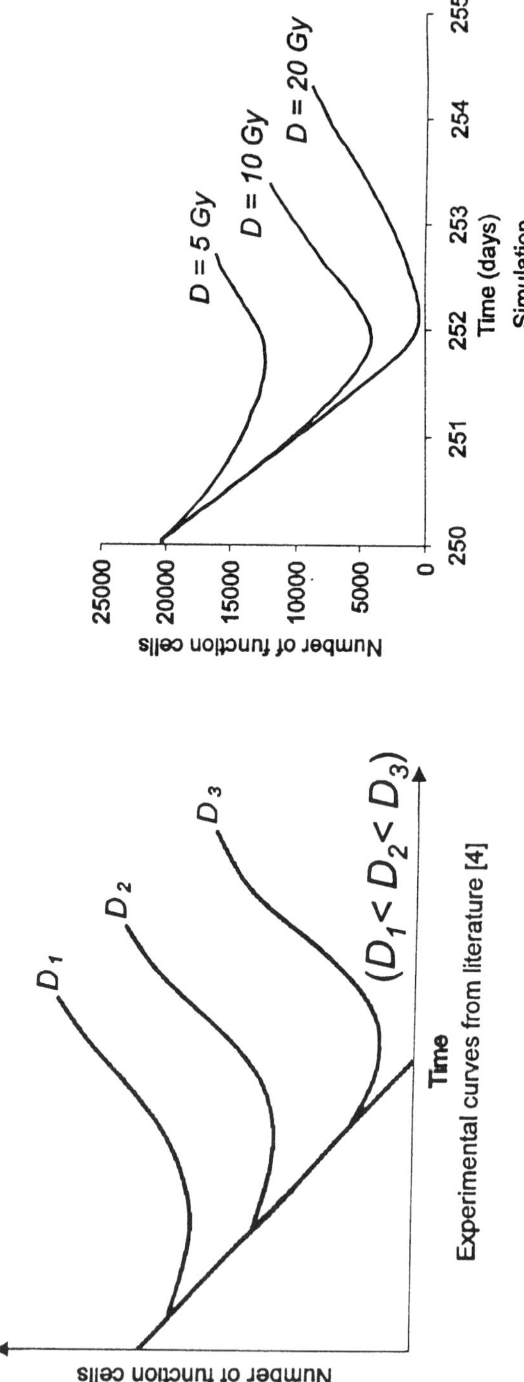

FIGURE 4.10 Cell number of function cells of the epidermis of the mouse after a single-dose irradiation at $T = 250$ days.

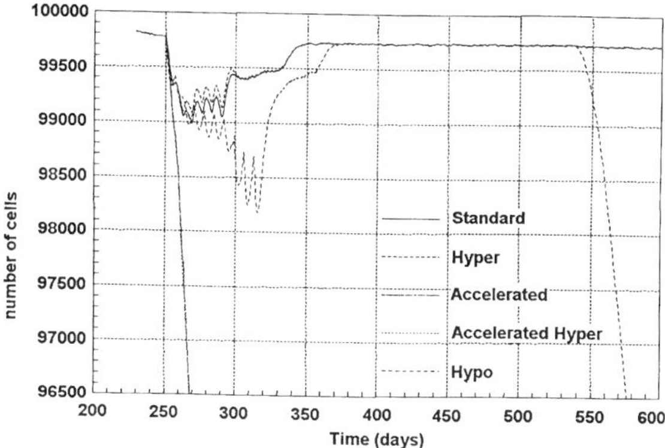

FIGURE 4.11 Irradiation of the lung parenchym of the mouse with five different fractionation schemes (see Table 4.1).

4.7 OUTLOOK

In this chapter we have seen that computer models are obviously essential simplifications of the true complexity of real life. So, much work remains to be done. Some promising avenues of future research include

- Giving stimuli to perform new biological experiments for measuring unknown data and parameters;
- Considering facts which had to be neglected so far (formation of metastases, immunologic reactions, drug resistance, heterogeneity);
- Extending the cell space which is at present limited to about 1 mm^3 tissue volume;
- Considering growth factors in hematopoietic cell renewal models;
- Generating a more realistic initial configuration of a tumor by combining image processing techniques, computer tomography (CT), magnetic resonance tomography (MR), and positron emission tomography (PET), with our predictive models describing tumor growth;
- Improving cancer treatment strategies by studying alternative schedules like continuous irradiation with low doses, administration of chemotherapeutic drugs, and/or combined therapy modalities, which are gaining an increasing importance in the treatment regimen of human tumor diseases. In this case, priority should be given to considering the cell metabolism and the nonlinear dependences of the metabolic variables;
- Comparing systematically different treatment methods and protocols;
- Partially substituting long and expensive biological test series with simulation experiments;
- Constructing computer models at the gene-regulation level by introducing methods from computer science, e.g., Petri networks (event-oriented asynchronous switch networks).

An importanct precondition (*conditio sine qua non*) of all modeling activities is the stepwise reduction of antipathy against the systematic modeling approach which is above all created by scientists predominantly working empirically. The authors sincerely hope that this chapter may be a component which inspires more confidence in modeling.

REFERENCES

1. Tautu, P., Mathematical models in oncology: a bird's-eye view, *Z. Krebsforsch.*, 91, 223–235, 1978.
2. Cherruault, Y., *Mathematical Modelling in Biomedicine,* D. Reidel Publishing Company, Dordrecht, 1986.
3. Thompson, J. R. and Brown, B. W., Eds., *Cancer Modeling,* Marcel Dekker, New York, 1987.
4. Wheldon, T., *Mathematical Models in Cancer Research,* Adam Hilger, Bristol, 1988.
5. Sandblad, B. and Meinzer, H. P., Modelling and simulation of complex control structures in cell biology, *Math. Inform. Med.,* 31, 36–43, 1992.
6. Murray, J. D., *Mathematical Biology,* Springer-Verlag, Berlin, 1993.
7. Drasdo, D., *Monte-Carlo-Simulationen in zwei Dimensionen zur Beschreibung von Wachstumskinetik und Strukturbildungsphänomenen in Zellpopulationen,* Verlag Shaker, Aachen, 1993.
8. Levin, S. A., Ed., *Frontiers in Mathematical Biology,* Springer-Verlag, Berlin, 1994.
9. Rowe, G., *Theoretical Models in Biology,* Clarendon Press, Oxford, 1994.
10. Mosekilde, E. and Mouritsen, O. G., Eds., *Modelling the Dynamics of Biological Systems,* Springer-Verlag, Berlin, 1995.
11. Solyanik, G. I., Berezetskaya, N. M., Bulkiewicz, R. I., and Kulik, G.I., Different growth patterns of a cancer cell population as a function of its starting growth characteristics: analysis by mathematical modelling, *Cell Prolif.,* 28, 263–278, 1995.
12. Denekamp, J., *Cellkinetics and Cancer Therapy,* C. C. Thomas, Springfield, IL, 1982.
13. Bannasch, P., Ed., *Cancer Therapy, New Trends,* Springer-Verlag, Berlin, 1989.
14. Horwich, A., Ed., *Oncology — A Multidisciplinary Textbook,* Chapman & Hall, London, 1995.
15. DeVita, V. T., Hellman, S., and Rosenberg, S. A., *Important Advances in Oncology 1995,* J.B. Lippincott, Philadelphia, 1995.
16. Düchting, W., Krebs, ein instabiler Regelkreis, Versuch einer Systemanalyse, *Kybernetik* 5, 70–77, 1968.
17. Bossel, H., *Modellbildung und Simulation,* Vieweg Verlag, Braunschweig, 1992.
18. Hogeweg, P., Simulating the growth of cellular forms, *Simulation,* 90–96, 1978.
19. Meinhard, H., *Models of Biological Pattern Formation,* Academic Press, London, 1982.
20. Keen, R. E. and Spain, J. D., *Computer Simulation in Biology,* Wiley-Liss, New York, 1992.
21. Anderson, J. G. and Katzper, M., Eds., *Simulation in the Health Sciences,* The Society for Computer Simulation, San Diego, 1994.
22. Düchting, W., Tumor growth simulation, *Comput. Graphics,* 3/4, 505–508, 1990.
23. Durand, R. E., Multicell spheroids as a model for cell kinetics studies, *Cell Tissue Kinet.,* 23, 141–159, 1990.
24. Ginsberg, T., Modellierung und Simulation der Proliferationsregulation und Strahlentherapie normaler und maligner Gewebe, Dissertation, Siegen, 1995.

25. Düchting, W., Ulmer, W., Ginsberg, T., and Saile, C., Radiogenic responses of normal tissue induced by fractionated irradiation — a simulation study. I. Acute effects, *Strahlenther. Onkol.,* 171, 460–467, 1995.
26. Potten, C. S., *Stem Cells,* Churchill Livingstone, Edingburgh, 1983.
27. Düchting, W. and Vogelsaenger, T., Three-dimensional pattern generation applied to spheroidal tumor growth in a nutrient medium, *Int. J. Bio-Med. Comput.,* 12, 377–392, 1981.
28. Düchting, W. and Dehl, G., Spatial structure of tumor growth: a simulation study, *IEEE Trans. Syst. Man Cybern.,* 10(6), 292–296, 1980.
29. Vogelsaenger, T., Modellbildung und Simulation von Regelungsmechanismen wachsender Blutgefäßstrukturen in normalen Geweben und malignen Tumoren, Dissertation, Siegen, 1986.
30. Düchting, W., Computer models applied to cancer research, in *Modeling and Control of Systems,* Blaquiere, A., Ed., Springer-Verlag, Berlin, 1988, 397–411.
31. Chabner, B. A. and Collins, J. M., *Cancer Chemotherapy,* J. B. Lippincott, Philadelphia, 1990.
32. Armitage, J. O. and Antman, K. H., Eds., *High-Dose Cancer Therapy,* Williams & Wilkins, Baltimore, 1992.
33. Eisen, M., *Mathematical Models in Cell Biology and Cancer Chemotherapy,* Springer-Verlag, Berlin, 1979.
34. Knolle, H., *Cell Kinetic Modelling and the Chemotherapy of Cancer,* Springer-Verlag, Berlin, 1989.
35. Kohn, K. W. and Ewig, R. A., DNA cross-linkings and cytotoxicity of normal and transformed cells, *Cancer Res.,* 38, 3197–3204, 1978.
36. Ulmer, W., Dose-effect relationships and 31P-NMR-spectroscopy of L1210 cells (monolayers) and 9L glioma cells (monolayers and tumor spheroids) treated with activated isophosphamide, adriamyacin, epirubicin and 6MeV electrons, *Strahlenther. Onkol.,* 167, 484–493, 1991.
37. Ulmer, W., In vivo investigations on murine C3H mammary adenocarcinoma treated with cis-platinum, adriamyacin, epirubicin, activated cyclophosphamide, activated isophosphamide and irradiation with 10 MeV electrons — verification of the therapeutic results with 31P-NMR-spectroscopy. *Strahlenther. Onkol.,* 167, 553–560, 1991.
38. Thames, H. D. and Hendry, J. H., *Fractionation in Radiotherapy,* Taylor & Francis, London, 1987.
39. Düchting, W., Ulmer, W., Lehrig, R., Ginsberg, T., and Dedeleit, E., Computer simulation and modeling of tumor spheroid growth and their relevance for optimization of fractioned radiotherapy, *Strahlenther. Onkol.,* 168, 354–360, 1992.
40. Swan, G. M., *Optimization of Human Cancer Radiotherapy,* Springer-Verlag, Berlin, 1981.
41. Perez, C. A. and Brady, L. W., *Principles of Radiation Oncology,* J. B. Lippincott, Philadelphia, 1989.
42. Steel, G. G., Adams, G. E., and Horwich, A., Eds., *The Biological Basis of Radiotherapy,* Elsevier, Amsterdam, 1989.
43. Withers, H. R., Some changes in concepts of dose fractionation over 20 years, *Front. Radiat. Ther. Oncol.,* 22, 1–13, 1988.
44. Freyer, J. P., Spheroids in radiobiology research, in *Spheroid Culture in Cancer Research,* Bjerkvig, R., Ed., CRC Press, Boca Raton, FL, 1992, 217–275.
45. Beck-Bornholdt, H. P., Ed., *Current Topics in Clinical Radiobiology of Tumors,* Springer-Verlag, Berlin, 1993.

46. Kocher, M. and Treuer, H., Reoxygenation of hypoxic cells by tumor shrinkage during irradiation, *Strahlenther. Onkol.,* 171, 219–230, 1995.
47. Düchting, W., Ginsberg, T., and Ulmer, W., Modeling of radiogenic responses induced by fractionated irradiation in malignant and normal tissue, *STEM CELLS,* 13 (Suppl. I), 301–306, 1995.
48. Düchting, W., Ulmer, W., Ginsberg, T., Kikhounga-N'Got, O., and Saile, C., Radiogenic responses of normal cells induced by fractionated irradiation — a simulation study, Part II. Late responses, *Strahlenther. Onkol.,* 171, 525–533, 1995.

5 A Model of HIV Infection and Its Blockers: From Complex to Simple

John L. Spouge

CONTENTS

5.1 INTRODUCTION

"The purpose of computing is insight, not numbers."

R. W. Hamming

An anonymous reviewer of one of my earliest papers[1] referred to the above quotation as "Hamming's classic dictum." The dictum originated in a different era of computing but holds true even today, with only a few qualifications. The truth of Hamming's dictum persists because scientific knowledge is usually discovered by one human being, who then is faced with communicating the results to somebody else. Nowadays, a user-friendly "expert" computer program may be able to bring one person's knowledge to bear directly on somebody else's data, but such programs are still laborious to write. Thus, unless a complex insight is extremely important, it must be simplified if you want to communicate it effectively. Such, I believe, is the essence of Hamming's dictum.

This chapter chronicles the progressive simplification of a model of HIV (human immunodeficiency virus) infection. I am hoping, probably not in vain, that in a volume devoted to the computer modeling of complex biological systems, this chapter's celebration of simplicity may be unusual enough to be interesting. The biology presented here is deliberately minimal to emphasize the modeling process, and the reader is referred to the references for more inclusive, and therefore more exact, biological presentations. Also, I have organized this article historically, so a discriminating reader is advised to skim the complexities of the initial mathematical model in favor of the simpler models and biological insights that appear later. The models presented are quite general, and they have biological applications other than HIV, as Section 5.5 suggests.

The initial mathematical model was motivated by conflicting results from test tube assays of an HIV blocker, *sCD4* (soluble CD4).[2-6] As with any other virus, successful infection requires HIV to attach, penetrate, uncoat, and replicate within a host cell.[7] This chapter focuses solely on the attachment step, which is mediated by viral attachment proteins, present in multiple copies and unique to each virus type. In HIV, the primary viral attachment protein is *gp120* (120-kD glycoprotein).[8,9] The attachment step usually requires the attachment protein to bind to a specific receptor site on target cells.[10] The main receptor for gp120 attachment is called *CD4*. The CD4 molecule is found on helper T lymphocytes, macrophages, and other immune cells, and makes them targets for infection by HIV *virions* (viral particles).

Historically, this suggested that soluble forms of the CD4 cell receptor (sCD4) might be able to complex with gp120 on the HIV surface, thereby blinding HIV virions to target cells and blocking viral attachment. The test tube assays mentioned above were intended to assess the therapeutic potential of sCD4, and the mathematical model presented in Section 5.2[11-13] was intended to give a theoretical framework for interpreting the discrepancies among the assay results.

5.2 THE LAYNE–SPOUGE–DEMBO (LSD) MODEL

Generically, the test tube assays of interest[14] examine *antiadsorptives*, blockers of viral attachment.[15] Such assays have three phases: preincubation, incubation, and amplification. During the optional *preincubation phase* (0 to 24 h), HIV virus is left to stand for some length of time, usually at room temperature. During the preincubation phase, some strains of HIV *shed* individual gp120 molecules on their surface;[16-18] a whole virion can also undergo *nonspecific inactivation*, completely losing its infectiousness spontaneously.[19] A preincubation phase can therefore be used to determine inactivation rates. Next, at the start of the *incubation phase* (1 to 2 h), target cells are mixed with a viral suspension, permitting virions to diffuse and attach to target cells. The antiadsorptive effectiveness of sCD4 can be measured by relative infection with and without sCD4 present. At the end of the incubation phase, a centrifuge spins the target cells down and separates them from the viral suspension. An *amplification phase* then amplifies cellular infection into a measurable signal.

Note that besides a direct antiadsorptive effect, high concentrations of sCD4 also accelerate gp120 shedding.[20-22] Since the sCD4 concentrations in our experiments were too low to accelerate shedding,[23] and since sCD4-accelerated shedding would unnecessarily divert the exposition away from modeling, this chapter ignores it.

To develop a computer model, consider a cell culture infected with a particular strain of HIV, and a viral suspension extracted from this culture. A suspension can be "optimized"[23] so that it contains a *cohort* of virions, all born at approximately the same time. For mathematical convenience, all virions in a cohort are assumed identical at birth, with a fixed number N of gp120 molecules on their surface. A virion remains *live* in the suspension if it has neither infected a cell nor been killed nonspecifically.

Equations 5.1 through 5.4 below constitute our original *LSD model*,[11,12] a model of the processes occurring when HIV infects helper T lymphocytes and sCD4 blocks HIV.

$$\frac{dI}{dT} = k_l LF \tag{5.1}$$

$$\frac{dV}{dT} = -k_l LF - k_n V \tag{5.2}$$

$$\frac{dF}{dT} = -k_f BF + k_r C - (k_s + k_n)F - k_l LF\left[1 + (N-1)\frac{F}{NV}\right] \tag{5.3}$$

$$\frac{dC}{dT} = k_f BF - k_r C - (k_s + k_n)C - k_l LF\left[(N-1)\frac{C}{NV}\right] \tag{5.4}$$

The independent variable in the LSD model is the time T. There are four dependent variables, one for each equation: I is the number of infected cells, V is the number of live virions, F is the number of *f*ree gp120 molecules on live virions, and C is the number of gp120 molecules on live virions *c*omplexed with sCD4. Lymphocyte and sCD4 (*b*locker) concentrations are denoted by L and B. In most experiments, the viral concentration is negligible compared with the other assay components, so L and B remain approximately constant. The quantities k_l, k_n, k_s, k_f, and k_r are rate constants, defined below.

The LSD model in Equations 5.1 through 5.4 makes an *equivalent site approximation*, a well-established approximation from polymer chemistry.[24] The equivalent site approximation asserts that each kinetic process treats gp120 molecules symmetrically. Thus, only the numbers F and C appear in Equations 5.1 through 5.4: details about the individual gp120 molecules on live virions, except for their blocking status, are irrelevant. In particular, Equation 5.1 implies that the rate of viral attachment is proportional to the number of free gp120s. Ultimately, the equivalent site approximation was justified by its agreement with experimental results.

Under the equivalent site approximation, the terms $k_l LF$ and $k_n V$ give the rates at which live viral particles are lost due to *l*ymphocyte infection and *n*onspecific killing, respectively. The terms $k_f BF$ and $k_r C$ are *f*orward and *r*everse rates for formation and dissociation of gp120–CD4 complexes. The terms $(k_s + k_n)F$ and $(k_s + k_n)C$ are the rates of loss of free and complexed gp120 from live virions due to the effects of spontaneous *s*hedding of individual gp120 molecules from live virions (k_s) and *n*onspecific inactivation of whole live virions (k_n). Finally, the terms $k_l LF[1 + (N - 1)F/NV]$ and $k_l LF[(N - 1)C/NV]$ are the rates of loss of free and complexed gp120 from live virions due to infectious events.

Now, imagine using Equations 5.1 through 5.4 to model an antiadsorptive assay. Let $V = V_0$ virions be preincubated from time $T = -T_p$ to time $T = 0$. During the preincubation phase, because virions do not mix with either lymphocytes or sCD4, $L = B = 0$. Analytic solution of Equations 5.1 through 5.4 from time $T = -T_p$ to time $T = 0$ with $L = B = 0$ yields $I = 0$, $V = V_0 \exp(-k_n T_p)$, $F = NV_0 \exp[-(k_n + k_s)T_p]$, and $C = 0$. Now, starting at time $T = 0$, incubate the virions with lymphocytes ($L > 0$), and possibly with sCD4 ($B \geq 0$). During the incubation phase, Equations 5.1 through 5.4 must be integrated numerically.

To apply the LSD model to actual assays, parameter ranges were estimated as follows[11,12]: $N \approx 80$, $k_n \leq 10^{-4}$ s^{-1}, $k_s \leq 10^{-4}$ s^{-1}, $k_f \leq 3 \times 10^{-12}$ cm^3 s^{-1}, and $k_r \leq 1.5$ s^{-1}. Although experiments vary, the sCD4 concentration B generally falls into the range 0 to 10^{15} molecules cm^{-3}; the lymphocyte concentration L, into the range 10^5 to 10^7 cells cm^{-3}; and the incubation time T, into the range 0 to 10^5 s. With these ranges in hand, we were able to simulate real assays and begin to simplify the LSD model.

5.3 SIMPLIFYING THE LSD MODEL

The LSD model in Equations 5.1 through 5.4 postulates a law governing viral infection, and like any other scientific law, it needs to be verified (see Reference 25, p. 156). To generate testable predictions, the model needs to be simplified, and its facets examined: "…the constructs or inventions that you make must be of such a kind that the consequences that you compute are comparable with experiment…" (Reference 25, p. 164) A mathematical model can generate two types of testable predictions: qualitative, possibly from observation of computer experiments; or quantitative, from mathematical analysis.

Unlike a laboratory experiment, which generates data, a computer experiment only generates *pseudodata*. (Experimenters enjoy this term, probably because it announces to them that your data are only pale imitations of theirs.) The pseudodata from Equations 5.1 through 5.4 indicated that sCD4 blocking becomes less effective as the cell concentration is increased.[11] This observation was easily explained.

Consider an antiadsorptive assay from the perspective of a specific random virion. At low lymphocyte concentrations, the virion has at most a single chance to attach and infect before inactivating. The presence of sCD4 can perturb this single event, reducing the number of infections by one. In contrast, at high lymphocyte concentrations, a virion has several chances to attach before inactivating. Because sCD4 is a reversible antiadsorptive, it must interrupt several independent attachment opportunities to prevent the virion from attaching and infecting. Thus, reversible antiadsorptives like sCD4 are predicted to be less effective at high lymphocyte concentrations. This theoretical prediction[11-13] was later confirmed, both in the test tube[23] and in clinical trials.[26,27]

Alerted to the importance of low lymphocyte concentrations, we generated pseudodata at these concentrations. Infection there was proportional to lymphocyte concentration: if only a few virions are given an attachment opportunity, doubling the lymphocyte concentration doubles the number of virions with an attachment opportunity, and doubles infection. Conceptually, we therefore distinguished between *unsaturated* assay conditions (in which only a few virions have an attachment opportunity) and *saturated* assay conditions, where a virion is "saturated" with attachment opportunities. Operationally, unsaturated and saturated assays could be distinguished by seeing whether or not infection is proportional to lymphocyte concentration.[14]

Mathematical analysis showed that over the full range of physically relevant parameters, viral gp120 and sCD4 reach equilibrium rapidly on the timescale required for target cell infection. This theoretical prediction and its consequences agreed with our data,[14,23] although its validity may not be universal.[28] In any case, it suggests a quasi-steady-state approximation, $k_f BF \approx k_r C$. Let $G = F + C$ be the total number of gp120 molecules on live virions, $K_{\text{assoc}} = k_f k_r^{-1}$ be the sCD4–gp120 association constant, and $K_d = K_{\text{assoc}}^{-1}$ be the corresponding dissociation constant. In an unsaturated assay ($L \approx 0$), infection is only a perturbation on the other processes ($k_l LF \approx 0$), so Equation 5.1 gives

$$\frac{I(B=0)}{I(B>0)} \approx \frac{k_l L \int_0^T G dt}{k_l L \int_0^T F dt} = \frac{\int_0^T (F+C) dt}{\int_0^T F dt} \approx 1 + \frac{k_f}{k_r} B = 1 + K_{\text{assoc}} B \tag{5.5}$$

An *inverse infection plot* plotting the *blocking ratio* $I(B = 0)/I(B > 0)$ against the sCD4 concentration B[11] should therefore yield a straight line whose slope equals the association constant K_{assoc}. Thus, a *chemical* equilibrium constant K_{assoc} can be measured in a *biological* assay.

The validity of Equation 5.5 may extend to any antiadsorptive based on a soluble receptor, e.g., soluble poliovirus receptor.[29,30] So far, however, Equation 5.5 has only been tested for sCD4. For sCD4, the blocking ratio for three different strains of HIV showed remarkably good experimental agreement with Equation 5.5, but only in the range $0 \leq B \leq K_d$.[14] At concentrations $B > K_d$, sCD4 blocking exceeded the prediction in Equation 5.5. Other data made cooperative blocking by sCD4 molecules unlikely, so the simplest explanation was that HIV infection requires cooperation between gp120 molecules, and that this cooperation became impaired when our experiments blocked more than 50% of the viral gp120. Thus, HIV infection appears to require a critical density of gp120 molecules on the viral surface.

Equation 5.5 also predicts that the preincubation time T_p does not influence sCD4 blocking. When the preincubation time T_p was varied, this prediction was confirmed up to $T_p \approx 7$ h, after which sCD4 blocking again exceeded the prediction in Equation 5.5. This observation could again be consistent with a critical density hypothesis, as follows. Before the preincubation phase, viral gp120 exceeds the critical density. As T_p increases, virions shed their gp120 and approach the critical gp120 density at $T_p \approx 7$ h. The sCD4 then becomes more effective than Equation 5.5 predicts, because it can push the free gp120 below the critical density.

5.4 MULTIPLICITY OF ATTACHMENT

This section sharpens "saturation" into a quantitative concept, *multiplicity of attachment*. Certain concrete considerations motivated the sharpening.

Because most of the viral burden in established HIV infection resides in lymph nodes,[31] antiviral therapy must be effective under lymph node conditions. Helper T lymphocyte concentrations in the lymph node are about 10^8 cells ml^{-1}, but lymphocyte concentrations in the test tube only reach about 10^6 cells ml^{-1}, because toxic cellular metabolites cannot be cleared there. HIV researchers are therefore forced to extrapolate test tube blocking to the lymph node, from 10^6 cells ml^{-1} to 10^8 cells ml^{-1}.

The theoretical prediction, that reversible antiadsorptives like sCD4 should be less effective at high lymphocyte concentrations, is only qualitative (saturated/unsaturated).[11-13] Despite this discouraging prediction, clinical trials of sCD4 costing many millions of dollars proceeded anyway and failed. Only one patient showed any therapeutic benefit, and that, only at the highest sCD4 dose attempted.[26,27] In retrospect, a *quantitative* prediction about the sCD4 concentrations required for therapy might have been extremely useful to clinicians.

To quantify the concept of saturation, consider for a moment the virological concept of MOI (multiplicity of infection). Common usage has muddied its meaning, but originally the MOI in an assay was defined as the average number of virions infecting a random target cell.[32] Virologists usually run their assays at a low MOI (<0.1) to ensure that very few cells are infected by more than one virion (<0.005 under Poisson statistics). At low MOI, a count of the infected cells precisely reflects the number of successfully infecting virions.

A low MOI is also required for assays of a viral blocker to be reproducible. Otherwise, at high MOI, lymphocytes may be infected by several virions, and the multiple infections may mask viral blocking. For symmetry, since MOI is a quantity associated with target cells, consider the virions in an assay now.

Prevailing assay conditions give any specific, random virion certain *attachment opportunities*, whereby the virion can attach, thereby committing itself irreversibly to a single lymphocyte. Let us define the virion's *multiplicity of attachment* (MOA) in any time interval of interest (e.g., the incubation time in an assay, or the viral infectious lifetime in a lymph node) to be the average number of attachment opportunities that must be blocked to keep the random virion in suspension.

MOA, which quantifies the concept of saturation, can be calculated explicitly as follows. Mathematically, attachment opportunities can be modeled as a Poisson process[33] of the general type.[34] In any particular assay run, and for any random virion, the actual number of attachment opportunities (that must be blocked in any particular time interval to keep the virion in suspension) is therefore Poisson distributed, with the mean equaling the MOA. If m denotes the MOA and a denotes the proportion of virions attaching,

$$a = 1 - e^{-m} \tag{5.6}$$

Let k_l be the rate of viral attachment; L, the lymphocyte concentration; and T, the virus cell incubation time. Then each virion has on average

$$m = \int_0^T k_1 L dT = k_l LT \tag{5.7}$$

attachment opportunities, where the simpler right-hand expression holds if the rate of viral attachment k_l and the cell concentration L both remain constant. The simpler expression often holds in practice, although certain experimental formats do invalidate it.[15]

In many assays, and perhaps in lymph nodes as well, viral infection is proportional to viral attachment. Equations 5.6 and 5.7 and then imply

$$I \propto 1 - e^{-m} = 1 - \exp\left(-\int_0^T k_l L dT\right) \tag{5.8}$$

At low MOA (e.g., $m < 0.1$) and constant lymphocyte concentration L, Equation 5.8 yields

$$I \propto 1 - e^{-m} \approx m = \int_0^T k_l L dT \propto \int_0^T k_l dT \tag{5.9}$$

Equation 5.5 can now be derived from Equation 5.9 without the full LSD model. The equivalent site approximation ("the rate of viral attachment is proportional to the number of free gp120s"), followed by Equation 5.9, yields

$$k_l(B=0) \propto G = F + C \Rightarrow I(B=0) \propto \int_0^T (F+C) dT \tag{5.10}$$

for $B = 0$, and

$$k_l(B>0) \propto F \Rightarrow I(B>0) \propto \int_0^T F dT \tag{5.11}$$

for $B > 0$, from which Equation 5.5 follows easily.

Although saturation was known to depend on cell concentration, Equation 5.7 explicitly shows that it and its quantitative counterpart, MOA, depend on incubation time as well. In fact, Equation 5.8 predicts that if an antiadsorptive only decreases k_l, it may have little effect on the infection I, if the MOA m was already large. Just as a high MOI masks viral blocking in general, a high MOA can mask blocking by some antiadsorptives.

The double proportionality in the right-hand expression of Equation 5.7, if applicable, also greatly facilitates comparisons: e.g., increasing incubation time ten-fold and cell concentration 100-fold increases a virion's attachment opportunities 1000-fold, as in the following example. HIV-1_{HXB3} (i.e., the HXB3 strain of HIV-1) inactivates spontaneously with a half-life of $\approx$10 h,[23] although a typical test tube incubation takes only $\approx$1 h. If the maximum lifetime of an HIV virion in a lymph node is $\approx$10 h, going from the test tube to the lymph node increases its incubation time ten-fold. In addition, test tube concentrations of helper T lymphocytes are usually $\approx 10^6$ cells ml^{-1}, whereas lymph node concentrations are $\approx 10^8$ cells ml^{-1}. The attachment rate for the MN strain of HIV-1 is $k_l = 10^{-5.85\pm0.06}$ ml h^{-1}, suggesting that to prevent infection, an antiadsorptive must block each HIV-1_{MN} virion about once in a typical test tube incubation ($m = k_l LT \approx 1$)[15], whereas it must block about a 1000 times in a lymph node ($m = k_l T \approx 1000$). Under these circumstances, neutralizing a single virion in a lymph node by blocking its attachment opportunities may be equivalent to the perfect neutralization of not tens, not hundreds, but thousands of equivalent virions in the test tube. It is hardly surprising that clinical trials of sCD4 failed.

Other explanations for the clinical failure have been advanced.[35-38] When Equations 5.6 and 5.7 and were used to extrapolate test tube data to lymph node conditions[15]; however, they predicted that only sCD4 concentrations exceeding the strain-specific gp120–sCD4 dissociation constant K_d would have a significant effect on lymphoid HIV, in at least semiquantitative agreement with the clinical findings. By accounting systematically for the effects of cell concentration and incubation time, the test tube studies, when properly interpreted, predicted the clinical results remarkably well.

5.5 DISCUSSION

Computer modeling of complex biological phenomena is an interdisciplinary enterprise that eventually involves communication among scientists with different backgrounds. This chapter has emphasized the necessity of simplifying computer results, to verify them as scientific theories and to communicate them across disciplines.

Without this simplification, scientists in recipient fields cannot apply the computer results effectively. For example, the LSD computer model in Equations 5.1 through 5.4[11-13] is incomprehensible to most virologists, but it was later simplified to make its consequences presentable in the virological literature,[14,15,23] yielding a number of practical insights for vaccine development and antiviral therapy.

The LSD model gave birth to several simpler concepts, namely, the concepts of a critical number of gp120s (which remained qualitative), the equivalent site approximation (Equation 5.5), and the viral MOA (Equation 5.7). We (LSD) tried to quantify the critical number of gp120s, but our partial success (or partial failure, if you prefer) was instructive. After postulating the concept of a critical number, we developed a computer model using $N = 80$ coupled ordinary differential equations, with the ith equation representing HIV virions with i gp120s on their surface. The model was eventually abandoned because its excessive detail made mathematical analysis extremely difficult. Without analysis, the complexity of the model placed it beyond quantitative verification, and without the possibility of quantitative verification, any quantitative insight from the model could only be a matter of faith, belonging properly to the province of theology, and not of science. Thus, in the end, we were forced to compare crude, simplified models to the available data, deriving only rough estimates of the critical number (around 4 to 40 gp120s, depending on the method and data). The original complex model remained unpublished, because simpler verbal arguments (see Equation 5.5 *et seq.*) sufficed to summarize our insights, which were primarily qualitative anyway.

Simplification and approximation can therefore make a complex computer model easier to compare with a particular class of data. More subtly, however, it can also widen the applicability of the model, so the model can be compared with broader classes of data. For example, the LSD model applies to a pathogen with multiple receptors for its target cell, but only if the pathogen satisfies the equivalent site approximation ("the rate of pathogen attachment is proportional to the number of its free attachment proteins"). Many blood-borne pathogens (lentiviruses like HIV, Epstein–Barr virus, and malaria) have multiple receptors, but it is not known if they satisfy the equivalent site approximation. The much simplified theory of viral MOA is independent of the equivalent site approximation, however, so finding preexisting data to compare with Equations 5.6 through 5.9 was easy.[14,15,39-41]

Thus, in conclusion, computers may make exploring the consequences of complex models much easier, but the twin needs of effective communication and experimental verification still require that simple, quantitative insights be extracted from these models. Thus, Hamming's dictum is worth remembering, even today: "The purpose of computing is insight, not numbers."

ACKNOWLEDGMENTS

I would like to acknowledge my collaborators in this work, S. P. Layne, M. Dembo, P. L. Nara, and S.-C. Wu, M. J. Merges, and S. Conley.

REFERENCES

1. Spouge, J. L., Monte Carlo results for random coagulation, *J. Coll. Interface Sci.*, 107, 38, 1985.
2. Deen, K. C., McDougal, J. S., Inacker, R., Folena-Wasserman, G., Arthos, J., Rosenberg, J., Maddon, P. J., Axel, R., and Sweet, R. W., A soluble form of CD4 (T4) protein inhibits AIDS virus infection, *Nature* (London), 331, 82, 1988.
3. Fisher, R. A., Bertonis, J. M., Meier, W., Johnson, V. A., Costopoulos, D. S., Liu, T., Tizard, R., Walker, B. D., Hirsch, M. S., Schooley, R. T., and Flavell, R. A., HIV infection is blocked *in vitro* by recombinant soluble CD4, *Nature* (London), 331, 76, 1988.
4. Hussey, R. E., Richardson, N. E., Kowalski, M., Brown, M. R., Chang, H. C., Siliciano, R. F., Dorfman, T., Walker, B., Sodroski, J., and Reinherz, E. L., A soluble CD4 protein selectively inhibits HIV replication and syncytium formation, *Nature* (London), 331, 78, 1988.
5. Smith, D. H., Byrn, R. A., Marsters, S. A., Gregory, T., Groopman, J. E., and Capon, D. J., Blocking of HIV-1 infectivity by a soluble, secreted form of the CD4 antigen, *Science*, 238, 1704, 1987.
6. Traunecker, A., Luke, W., and Karjalainen, K., Soluble CD4 molecules neutralize human immunodeficiency virus type 1, *Nature* (London), 331, 84, 1988.
7. Bukrinskaya, A. G., Pentration of viral genetic material into host cell, *Adv. Virus Res.*, 27, 141, 1982.
8. Dalgleish, A. G., Beverley, P. C., Clapham, P. R., Crawford, D. H., Greaves, M. F., and Weiss, R. A., The CD4 (T4) antigen is an essential component of the receptor for the AIDS retrovirus, *Nature*, 312, 763, 1984.

9. Klatzmann, D., Champagne, E., Chamaret, S., Gruest, J., Guetard, D., Hercend, T., Gluckman, J. C., and Montagnier, L., T-lymphocyte T4 molecule behaves as the receptor for human retrovirus LAV, *Nature* (London), 312, 767, 1984.
10. Tardieu, M., Epstein, R. L., and Weiner, H. L., Interaction of viruses with cell surface receptors, *Int. Rev. Cytol.*, 80, 27, 1982.
11. Layne, S. P., Spouge, J. L., and Dembo, M., Quantifying the infectivity of human immunodeficiency virus, *Proc. Natl. Acad. Sci. U.S.A.*, 86, 4644, 1989.
12. Layne, S. P., Spouge, J. L., and Dembo, M., The kinetics of HIV infectivity, *Los Alamos Sci.*, 18, 91, 1989.
13. Spouge, J. L., Layne, S. P., and Dembo, M., Analytic results for quantifying HIV infectivity, *Bull. Math. Biol.*, 51, 715, 1989.
14. Layne, S. P., Merges, M. J., Dembo, M., Spouge, J. L., and Nara, P. L., HIV requires multiple gp120 molecules for CD4-mediated infection, *Nature* (London), 346, 277, 1990.
15. Spouge, J. L., Viral multiplicity of attachment and its implications for HIV therapies, *J. Virol.*, 68, 1782, 1994.
16. Gelderblom, H. R., Reupke, H., and Pauli, G., Loss of envelope antigens of HTLV-III/LAV, a factor in AIDS pathogenesis? [letter], *Lancet*, 2, 1016, 1985.
17. Kowalski, M., Potz, J., Basiripour, L., Dorfman, T., Goh, W. C., Terwilliger, E., Dayton, A., Rosen, C., Haseltine, W., and Sodroski, J., Functional regions of the envelope glycoprotein of human immunodeficiency virus type 1, *Science*, 237, 1351, 1987.
18. Gelderblom, H. R., Ozel, M., and Pauli, G., Morphogenesis and morphology of HIV. Structure-function relations, *Arch. Virol.*, 106, 1, 1989.
19. McDougal, J. S., Martin, L. S., Cort, S. P., Mozen, M., Heldebrant, C. M., and Evatt, B. L., Thermal inactivation of the acquired immunodeficiency syndrome virus, human T lymphotropic virus-III/lymphadenopathy-associated virus, with special reference to antihemophilic factor, *J. Clin. Invest.*, 76, 875, 1985.
20. Moore, J. P., McKeating, J. A., Weiss, R. A., and Sattentau, Q. J., Dissociation of gp120 from HIV-1 virions induced by soluble CD4, *Science*, 250, 1139, 1990.
21. McKeating, J. A., McKnight, A., and Moore, J. P., Differential loss of envelope glycoprotein gp120 from virions of human immunodeficiency virus type 1 isolates: effects on infectivity and neutralization, *J. Virol.*, 65, 852, 1991.
22. Moore, J. P., McKeating, J. A., Norton, W. A., and Sattentau, Q. J., Direct measurement of soluble CD4 binding to human immunodeficiency virus type 1 virions: gp120 dissociation and its implication for virus-cell binding and fusion reactions and their neutralization by soluble CD4, *J. Virol.*, 65, 1133, 1991.
23. Layne, S. P., Merges, M. J., Spouge, J. L., Dembo, M., and Nara, P. L., Blocking of human immunodeficiency virus infection depends on cell density and viral stock age, *J. Virol.*, 65, 3293, 1991.
24. Flory, P. J., Molecular size distributions in three dimensional polymers. I. Gelation, *J. Am. Chem. Soc.*, 63, 3083, 1941.
25. Feynman, R., *The Character of Physical Law*, M.I.T. Press, Cambridge, MA, 1967.
26. Schooley, R. T., Merigan, T. C., Gaut, P., Hirsch, M. S., Holodniy, M., Flynn, T., Liu, S., Byington, R. E., Henochowicz, S., Gubish, E., Spriggs, D., Kufe, D., Schindler, J., Dawson, A., Thomas, D., Hanson, D. G., Letwin, B., Liu, T., Gulinello, J., Kennedy, S., Fisher, R., and Ho, D. D., Recombinant soluble CD4 therapy in patients with the acquired immunodeficiency syndrome (AIDS) and AIDS-related complex. {A} phase I-II escalating dosage trial, *Ann. Intern. Med.*, 112, 247, 1990.

27. Kahn, J. O., Allan, J. D., Hodges, T. L., Kaplan, L. D., Arri, C. J., Fitch, H. F., Izu, A. E., Mordenti, J., Sherwin, J. E., Groopman, J. E., and Volberding, P. A., The safety and pharmacokinetics of recombinant soluble CD4 (rCD4) in subjects with the acquired immunodeficiency syndrome (AIDS) and AIDS-related complex. A phase 1 study, *Ann. Intern. Med.,* 112, 241, 1990.
28. Dimitrov, D. S., Hillman, K., Manischewitz, J., and Blumenthal, R., Kinetics of soluble CD4 binding to cells expressing human immunodeficiency virus type 1 envelope glycoprotein, *J. Virol.,* 66, 132, 1992.
29. Kaplan, G., Peters, D., and Racaniello, V. R., Poliovirus mutants resistant to neutralization with soluble cell receptors, *Science*, 250, 1596, 1990.
30. Kaplan, G., Freistadt, M. S., and Racaniello, V. R., Neutralization of poliovirus by cell receptors expressed in insect cells, *J. Virol.,* 64, 4697, 1990.
31. Pantaleo, G., Graziosi, C., and Fauci, A. S., New concepts in the immunopathogenesis of human immunodeficiency virus infection, *N. Engl. J. Med.*, 328, 327, 1993.
32. Ellis, E. L. and Delbrueck, M., The growth of bacteriophage, *J. Gen. Physiol.*, 22, 365, 1939.
33. Bailey, N. T. J., *The Elements of Stochastic Processes*, John Wiley and Sons, New York, 1964.
34. Doob, J. L., *Stochastic Processes*, John Wiley and Sons, New York, 1953.
35. Daar, E. S., Li, X. L., Moudgil, T., and Ho, D. D., High concentrations of recombinant soluble CD4 are required to neutralize primary human immunodeficiency virus type I isolates, *Proc. Natl. Acad. Sci. U.S.A.*, 87, 6574, 1990.
36. Kohler, H., Goudsmit, J., and Nara, P. L., Clonal dominance: cause for a limited and failing immune response to HIV-1 infection and vaccination, *J. Acquired Immune Defic. Syndr.*, 5, 1158, 1992.
37. Orloff, S. L., Kennedy, M. S., Belperron, A. A., Maddon, P. J., and McDougal, J. S., Two mechanisms of soluble CD4 (sCD4)-mediated inhibition of human immunodeficiency virus type 1 (HIV-1) infectivity and their relation to primary HIV-1 isolates with reduced sensitivity to sCD4, *J. Virol.,* 67, 1461, 1993.
38. Zolla-Pazner, S., Goudsmit, J., and Nara, P. L., Characteristics of human neutralizing antibodies derived from HIV-1 infected individuals, In *Seminars in Virology*, Vol. 3, W. B. Saunders, Philadelphia, 1992, 203.
39. Allison, A. C. and Valentine, R. C., Virus particle adsorption: III. Adsorption of viruses by cell monolayers and effects of some variables on adsorption, *Biochim. Biophys. Acta*, 40, 400, 1960.
40. Allison, A. C. and Valentine, R. C., Virus particle adsorption: II. Adsorption of vaccinia and fowl plague viruses to cells in suspension, *Biochim. Biophys. Acta*, 40, 393, 1960.
41. Mandel, B., Studies on the interaction of poliomyelities virus, antibody, and host cells in a tissue culture system, *Virology*, 6, 424, 1958.

6 Modeling HIV Infection Kinetics

Dimiter S. Dimitrov

CONTENTS

6.1 MATHEMATICAL MODELING AND OUR UNDERSTANDING OF VIRUS INFECTIONS

Several decades ago Delbruck[1,2] among others developed quantitative methodology, based on precise mathematical and physicochemical terms, for studying bacteriophages infecting bacterial cultures, which helped in fully evaluating important genetic concepts. Dulbecco[3] introduced the plaque assay to animal virology, which was a direct extension of studies involving bacteriophages. The concepts derived from these early studies, e.g., multiplicity of infection (MOI),[4] have been widely used for characterization of animal viruses. Later, mathematical models were developed to describe kinetics of virus adsorption based on concepts from colloid chemistry.[5,6]

While very helpful in understanding virus infection kinetics, these early models were limited to simple equations which allowed analytical solutions, or required techniques for derivation of relationships between system parameters without solving the equations, or described limiting cases of unknown general solutions. With the recent rapid development of computer technology, it is possible now to develop and analyze much more complex models. This leads to more opportunities for exploration and potentially deeper understanding of mechanisms of virus infections. A note of caution is, however, due related to the complexity of biological systems and the limitations of current mathematical tools for their description. In most cases, the numerical values of the parameters used to model complex systems are not known and their interrelationships are not rigorously proved in independent experiments. Thus, a clear distinction should be made between mathematical models which try to prove a concept but are not based on solid and comprehensive quantitative data and models which describe quantitatively experimental observations in well-defined systems and can predict results in agreement with data. Both types of models can generate important ideas for future experiments, help in the interpretation of experiments, and increase our understanding of virus infections, although the "simple" models, based on carefully derived, quantitative data, certainly have dominated in their contribution to the field of virology. The future may witness, however, how sophisticated computer-assisted modeling dramatically enhances our understanding of viral infections.

In the rest of this chapter, I will first briefly overview basic experimental observations and several mathematical models of human immunodeficiency virus (HIV) infection in humans. Then I will describe in more details our model of HIV infection kinetics in tissue cultures and its implications for understanding HIV infection kinetics. Finally, I will discuss estimations of kinetic parameters in patients, a model of "physiological" variations in virus concentrations, and the implications of modeling for therapy and vaccine development.

6.2 HIV INFECTION KINETICS: BASIC EXPERIMENTAL OBSERVATIONS

Most of our knowledge on HIV infection kinetics has been derived from studies of tissue cultures.[7,8] Recent developments in (1) accurate quantitation of HIV RNA and DNA by PCR and branched DNA techniques, (2) development of powerful inhibitors of HIV type 1 (HIV-1) replication, and (3) animal models of HIV infections have made available a wealth of information for HIV infection kinetics at the level of whole organisms. The complexity of animal organisms compared with tissue cultures involves but is not limited to (1) existence of an immune system that attempts to reduce virus load in a complex and not completely understood way, which is, however, clearly dependent on host- and virus-related factors; (2) different types of cells that can replicate virus at different rates and to a certain extent that are dependent on each other, particularly through a variety of cytokines; (3) existence of different compartments of virus replication, particularly lymphoid tissues, brain, and blood, where cell type and concentration, as well as levels of virus replication related to the state of cell activation, differ significantly; and (4) the history of virus infection, the type of infecting virus quasi species, and their evolution in the host could differ very significantly between individuals who are exposed to stochastic ill-controlled variations. In spite of the significant differences between tissue culture systems and whole organisms, a number of similarities can justify experimental

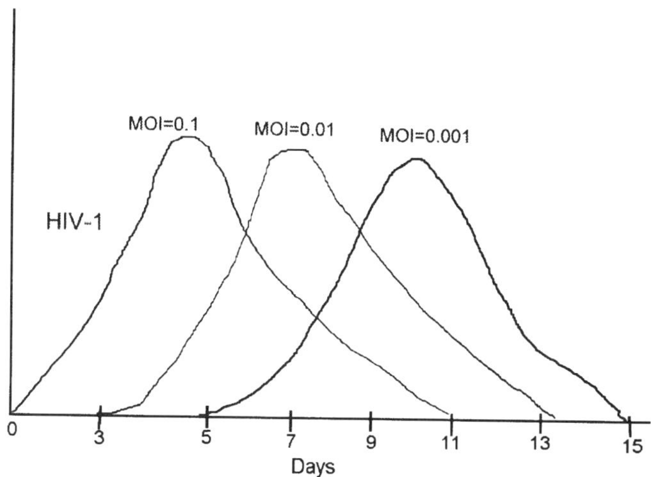

FIGURE 6.1 HIV-1 kinetics of infection in tissue cultures. A typical course of infection as a function of time at different initial MOI is shown. MOI is defined as the ratio of infectious virus units divided by the number of uninfected cells. Similar functional dependences are also valid for the number of infected cells as a function of time, while the number of uninfected CD4 cells declines as the number of infected cells rises (not shown).

studies with tissue cultures: (1) the virus life cycle contains essentially the same elements in tissue culture cells as in cells from infected hosts; (2) spread by a cell-free or cell-associated virus in hosts could be mimicked in tissue cultures; and (3) effects of a variety of molecules in tissue cultures, including antibodies, infection inhibitors, and stimulators, have their analogues during virus–host interactions. In the rest of this section, I will emphasize similarities between tissue cultures and infected hosts rather than differences.

All HIV infections are initiated either by cell-free or cell-associated (infected cells) virus (except in cases where under experimental conditions plasmid DNA, containing provirus, is used to transfect cells — that procedure is usually used to produce relatively homogeneous cloned virus but not to initiate new infections). From a kinetic point of view the significant difference between these two cases is that infected cells are much more effective than cell-free virus for three major reasons: (1) infected cells can produce many (on average 100 but in some cases up to 1000) infectious units (IU) of virus; (2) the virus is better protected inside cells (as a provirus DNA which is capable of expression upon activation) than as an extracellular RNA-containing virion, where the ratio of IU to virion particles can be as small as 1 to 10^4 to 10^5 and declines with time if the virus does not enter a target cell; and (3) transmission of virus from infected to uninfected cells can be very efficient especially in situations where cells are in close contact.

In tissue cultures of CD4 T cells, a typical course of infection includes an initial stage of an exponentially growing number of virions and infected cells reaching a peak followed by a decline until all cells die and virus production ceases (Figure 6.1). The time required to reach the infection peak increases with a decrease in the initial

MOI. Different variations of this basic infection kinetic pattern in tissue cultures include

1. Purified CD4 T cell from peripheral blood lymphocytes (PBLs) may not be readily infectable unless activated;
2. Many cells including CD4 T cells from PBLs do not die after infection; some of them survive, for example, by downmodulating the surface expression of CD4, others carry the HIV provirus but it is not active, and yet another subset of cells may become chronic producers of HIV albeit at low levels compared with acutely infected cells;
3. Macrophages and other non-T cells may show slow infection kinetics and a lower level of virion production which may be undetectable unless they are incubated with CD4 T cells which can pick up infectious virus due to the high efficiency of cell-to-cell transmission;
4. Culturing of continuous cell lines requires splitting of the cultures at regular interval (e.g., threefold every 2 days) to keep the cell concentration approximately constant at least through the initial stages of infection; that may lead to eradication of virus infection if the rate of virus spread is smaller than the rate of splitting, and no peak of viremia or even virus will be observed; and
5. In some cases, particularly when cultures of continuously growing cells are not split, virion concentration can rise and fall periodically.

In infected individuals the initial stages of infection before the appearance of immune response may well mimic tissue culture infections: the virus and infected cells rise very rapidly (and CD4 cells decline) until reaching a peak (typically within several weeks) after which they decline and CD4 cells increase in number to reach approximately preinfection levels within several months (Figure 6.2). This period of acute primary infection is followed by an asymptomatic period of variable length (in some cases up to 10 years or longer) where initially the virus is at very low levels and the number of uninfected CD4 cells is close to normal (about 1000/μl in blood). However, at some point during this asymptomatic period the number of virus and infected cells begins to rise accompanied by a decline in CD4 cells. This process continues until reaching the final stage of the infection characterized as acquired immunodeficiency syndrome (AIDS) where levels of viremia can reach values as high as millions of virions per milliliter of blood and CD4 cells can decline down to undetectable levels. This basic pattern of infection kinetics varies significantly between individuals, depending on a complex interplay between host and viral factors. It seems that virus replication is controlled primarily by cellular immune response, including cytotoxic T lymphocytes (CTLs), which could appear as early as a week after initiation of infection, while it may take a month or longer for production of neutralizing antibodies. Thus, while the peak of infection in tissue cultures is determined by the mere exhaustion of target infectable cells, it appears that in infected individuals the level of activity of the immune system is the major determinant of how rapidly the viremia will begin to decline.

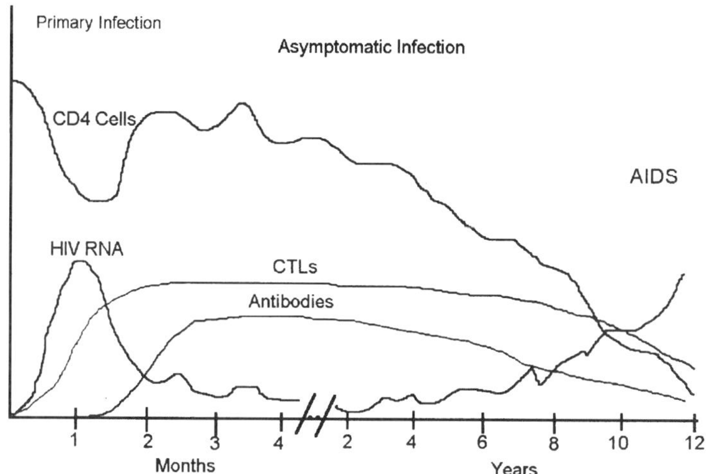

FIGURE 6.2 Progression of HIV-1 infection in patients following three major stages: primary acute infection, asymptomatic phase, and a final stage ending with AIDS. The number of virions is proportional to the number of productively infected cells and in an inverse correlation with the number of uninfected CD4 cells. A typical course of infection is shown. In infected individuals (nonprogressors) AIDS never occurs and in many of the nonprogressors virus load never rises after the asymptomatic period; in a few, however, it can be quite high.

Any model of HIV infection must predict these basic kinetic patterns or parts of them at reasonable values of the system parameters. In the following section, two models which attempt to describe the decline in CD4 cells during progression to AIDS and the peak of viremia during primary infections, respectively, are briefly reviewed. These theories are typical for mathematical modeling of HIV infection and represent approaches similar to those used for developing models of the infection kinetics in tissue cultures which are described later.

6.3 MODELS OF HIV INFECTION KINETICS: AN OVERVIEW

During the last decade a number of mathematical models have been developed to describe a variety of features of HIV-1 infection, including effects of the immune system and treatments with inhibiting agents.[9-22] While in the past most of the models attempted to model interactions of HIV with the immune system that are important for pathogenesis, during the last few years several models elaborated explicitly on description of HIV-1 infection kinetics without taking into account the immune response. This situation may occur either late in the infection process when the immune response has already deteriorated, or else, in the very beginning, before the appearance of the response. The two models described below attempt to analyze such cases.

Perelson et al.[16] developed an elaborate model that considers the kinetics of uninfected, latently and productively infected cells, and virus, in an attempt to predict the CD4 T cell decline solely on the basis of population dynamics. The model does not account for the immune response and assumes that cell deaths results from direct HIV-mediated killing. It predicts that T cell depletion is critically dependent on the burst size of virus infection — the number of IU, n, produced by one productively infected cell; in dependence on n the T cell depletion can occur within 1 to 3 years. However, the model predicted that the number of productively infected cells was relatively high — more than 10% of all CD4 T cells. To correct for this problem it was assumed that HIV may infect precursor cells, thus effectively decreasing the rate of CD4 cell regeneration. This led to a more realistic number of productively infected cells (between 1 in 100 and 1 in 1000). However, in this case the CD4 cell concentration was relatively high, higher than 500 cells/μl, while in many patients with AIDS it is below 200. This problem could be corrected if one assumes that n increases with the progression of the HIV disease, e.g., by switching from the so-called slow- to rapid-replicating viruses. The critical importance of n was also demonstrated in another fundamental aspect: below a critical value of n, a spreading infection cannot be sustained. Under such conditions the level of free virus monotonously decreased and was ultimately eliminated. The model also predicted that oscillations in viral load require parameters outside the region of biologically realistic parameters and that treatment with the antiviral drug AZT may lead to recovery if it reduces n below its critical value. While this model is oversimplified, it appears that it describes several characteristic features of the HIV infection, including the existence of a critical n and effects of impairments in CD4 cell regeneration on their decline, which may be critical for the development of the disease.

Phillips[22] developed a model specifically describing the acute primary infection. It predicted that the reduction of plasma virus load can be explained by population dynamics and it may not reflect an immune response to the virus. While the effects of population dynamics can lead to transient decreases and oscillations of virus produced in tissue culture systems,[23] it appears that the total concentration of virions (infectious and defective) predicted by the model[22] is unrealistically high.

As in the previously described model[16] a critical parameter embodied in the primary infection model[22] is n. The author uses a value of 100 for this parameter, referring to our previously published report,[24] where we estimated the number of infectious virions produced by one infected cell during a spreading virus infection. In our paper we emphasized that cell-to-cell spread is the dominant mechanism of virus transmission and the value we found (100) represents the number of infectious particle equivalents transmitted per cell. In contrast, the number of infectious virions released from a single cell and capable of initiating a cell-free virus infection is 100 to 1000 times lower than this value, reflecting the variable particle-to-infectivity ratios produced under different experimental conditions. Since the model[22] deals only with cell-free virus transmission and not with cell-to-cell HIV-1 spread, the calculated plasma virion concentration may not be correct.

Another major problem of the model[22] is the failure to distinguish between the small number of infectious virions in an "ocean" of defective virions. For example, when Phillips reports the rate of virion production or the initial HIV-1 inoculum, he assumes that all of the virions are infectious. His model then predicts that at the peak of the primary infection the virion concentration V is about 10^7/ml, equivalent to a value of 10^7 RNA copies/ml which agrees with previously published experimental data.[25] However, it is now recognized that both in tissue culture systems[24] and in infected individuals,[26] the ratio of defective virions to infectious virions is very high (in the range from 1000 to more than 10,000). With this correction, the model would predict that the plasma virus concentration at the peak of the primary infection would be approximately $10^7 \times 10^3 = 10^{10}$ RNA copies/ml, an unrealistic value.

This discrepancy cannot be resolved by simply renormalizing parameters because the number of productively infected CD4 cells during a primary HIV-1 infection is not known. For example, the virus concentration could be reduced by decreasing the proportion of activated cells. However, then the model[22] would not predict any significant decline in the number of CD4 cells following infection. The high viral loads can also be reduced by decreasing virus production to ten infectious virions per cell, which may be a reasonable value for the *in vivo* situation.[27] However, our computer simulations of the equations and parameters, described in Reference 22, revealed that lower virus production resulted in the eradication of the virus infection. When the infection rate constant was increased tenfold to compensate for decreased virus production and to allow for spreading infection, the number of virions was still several orders of magnitude larger than the experimentally observed values.

The unrealistically high viral loads predicted by Phillips's model can be resolved by allowing the immune response to remove a portion of the infectious virus progeny after a certain period of time and by taking into account the cell-to-cell spread of the virus. Thus, in addition to the experimental evidence indicating a correlation between the appearance of an immune response and decline in virus load,[28] a more careful examination of HIV-1 infection kinetics models also suggests that other mechanisms (acting independently or in cooperation with population dynamics effects) exist to remove virus during primary infections.

This detailed analysis of a mathematical model of HIV infection kinetics demonstrates how values of experimental parameters are critical for interpretation of results. Despite the criticism, the Phillips model is an important step in the development of mathematical models of HIV infection kinetics, showing how modeling can raise interesting ideas and view well-known phenomena from an entirely different angle.

6.4 HIV INFECTION KINETICS IN TISSUE CULTURES

As with many other viruses, most of our knowledge about the replication of HIV-1 has been derived from studies of infection in tissue cultures (for a review and references see References 7 and 8). Surprisingly, in spite of the enormous amount of work performed with tissue culture HIV-1 infections, they have not been studied theoretically as bacteriophages were in the late 1930s. An exception is the derivation of a semiempirical formula based on a model of the HIV-1 spreading in tissue cultures by subsequent cycles of infection.[24] This formula predicted an exponential growth of the virus and allowed the calculation of two important parameters of the viral infection: (1) the number of infecting virions produced by one cell in one cycle of infection and (2) the time required to complete one cycle of infection. The formula was also useful in determining the virus infectivity and how it was changed after treatment with inhibitors of HIV-1 replication. However, this formula was limited to the initial stages of viral growth and did not describe the later stages of infection. In addition, it did not allow consideration of a number of factors which affect the virus infection kinetics, including the relative role of cell-to-cell spread vs. infection by cell-free virus during the virus spread, the role of virus attachment, dead cells, etc. To account for those factors, a more-sophisticated mathematical model of HIV-1 infection kinetics in tissue cultures based on a system of differential equations similar to those already used to model the HIV-1 infection kinetics *in vivo*[16] was developed.[23] However, unlike previous models we take explicitly into account the cell-to-cell spread which in most cases seems to dominate tissue culture infections.[24,29] The fact that tissue cultures are inherently much simpler systems than the infections of whole organisms allowed us to use experimental parameters which are more accurately determined than those *in vivo*, derive conclusions which may have predictive power, and provide a better interpretation of the experimental data.

6.4.1 Cell-to-Cell Spread

The simplest system of differential equations which can still represent characteristic features of HIV-1 infection kinetics in tissue cultures by cell-to-cell transmission is the following[23]:

$$C' = -ki * I * C + g * C * \left[1 - (C - I)/Cm\right] \tag{6.1a}$$

$$I' = ki * I * C - m * I \tag{6.1b}$$

where C and I are concentrations of uninfected and infected cells, respectively, ki (2×10^{-6}/ml/day) is an infection rate constant; g (0.7/day) and m (0.3/day) are uninfected cell growth and infected cell death rate constants, respectively; Cm (2×10^6/ml) is the carrying capacity of healthy uninfected cells; $'$ and $*$ denote derivatives with respect to time and multiplication, respectively. The values in parenthesis denote typical (default) values used for simulations.

In a typical tissue culture experiment, most of the time the cells are not well mixed and make good contact with the neighbor cells. Therefore, it is not surprising that the cell-to-cell mode of virus transmission dominates.[29,30] However, the tissue culture cells are frequently split (usually every 2nd or 3rd day), and during the removal of part of the cells, all the cells are mixed. In addition, lymphocytes move if not aggregated. This partly justifies using Equations 6.1 which assume that the infected cells make frequent contacts with uninfected in proportion to their concentrations. In the rest of this section we will analyze numerical solutions of Equation 6.1 which demonstrate several characteristic features of a spreading infection in tissue cultures. For simplicity, we will assume that dead cells do not contribute significantly to slowing down healthy cell reproduction. Taking into account the dead cells does not change the qualitative features of the numerical solutions.

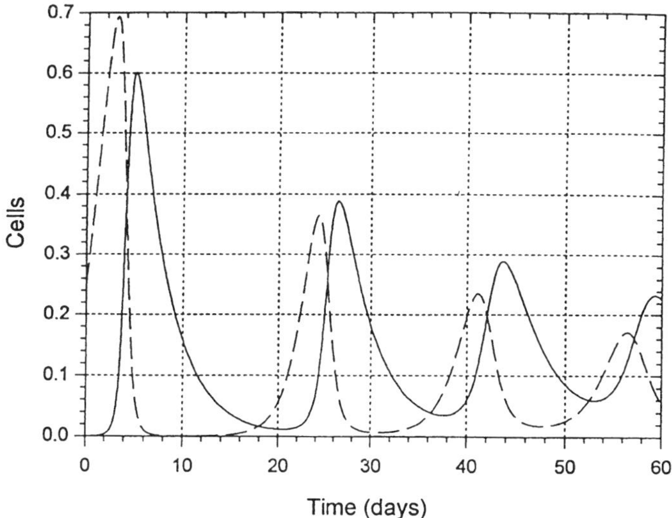

FIGURE 6.3 HIV-1 infection kinetics in tissue cultures by cell-to-cell spread of the virus as simulated by the solution of Equations 6.1 with parameters: $ki = 2 \times 10^{-6}$/ml/day, $g =$ 0.7/day, $m = 0.3$/day and $Cm = 2 \times 10^6$/ml. The dimensionless variables, the concentration of infected (solid lines) (I/Cm, initial value = 0.25) and uninfected (dashed lines) (C/Cm, initial value = 0.00025) cells, are shown as functions of the dimensional independent variable time (days). (From Spouge, J. L. et al., *Bull. Math. Biol.*, 138, 1, 1996.)

A basic parameter in Equations 6.1, which depends on the particular virus/cell system and to somewhat less extent on the conditions of culturing, is the infection rate constant, *ki*. We will determine the range of values of this parameter for typical tissue culture experiments by assuming that the cell-to-cell spread is the dominant mode of transmission and use experimental data for virus spread.[24] We have previously used the concept of an infection rate constant *k* defined as

$$dI/dt = kI$$

to describe the initial stages of a viral infection.[24] Then from Equations 6.1 one can estimate *ki* as k/C, where C is a characteristic average healthy cell concentration in the initial stages of viral infection, typically of the order of 10^6/ml. The measured infection rate constant *k* varied between 0.3 (typical for macrophages) to 1 to 1.5/day (typical for T cells). Therefore, one can assume that *ki* varies in the range from 0.3 to 2×10^{-6}/day/ml. The values of the other parameters, as shown as default values, represent those found for T cell lines, e.g., CEM cells. They also can vary in dependence on the particular culture conditions and the cell/virus system.

Figure 6.3 shows a numerical solution of Equations 6.1 obtained by using the computer program Scientist (MicroMath, Salt Lake City, Utah). It can be seen that initially the concentration of infected cells rises, reaches a maximum, and then begins to decline. Interestingly, however, after reaching a minimum the concentration of infected cells begins to rise again. The cycle repeats with a frequency on the order

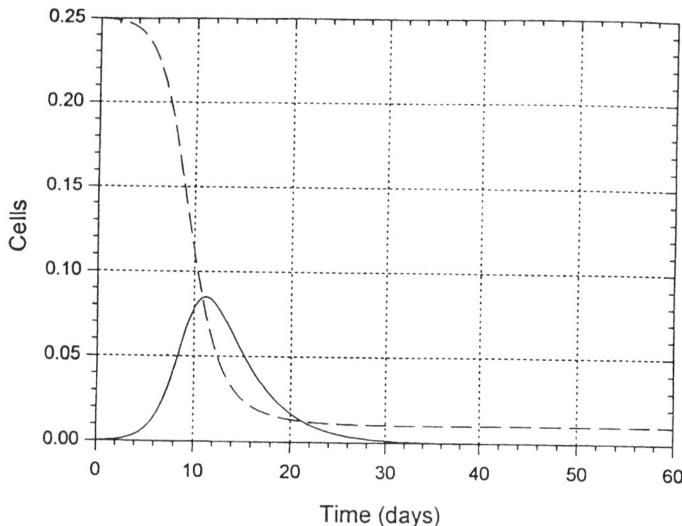

FIGURE 6.4 HIV-1 infection kinetics in tissue cultures by cell-to-cell spread of the virus as simulated by the solution of Equations 6.1 with added terms to account for splitting. The culture was split to match the cellular reproductive rate, $k_s = g$, ensuring zero net growth of uninfected cells. The initial conditions, parameters, and dimensionless variables are the same as in Figure 6.3. The dimensionless variables, the concentration of infected (solid lines), and uninfected (dashed lines) cells, are shown as functions of the dimensional independent variable time (days). (From Spouge, J. L. et al., *Bull. Math. Biol.*, 138, 1, 1996.)

of the square root of the product of the cell growth, g, and dead, m, rate constants. The height of the maximum decreases with each cycle in proportion to the value of the carrying capacity of healthy cells, Cm. This type of behavior is typical for the dynamics of predator–prey populations, described by a Lotka–Volterra type of equations.[31]

As mentioned above, during the maintenance of tissue cultures part of the healthy and infected cells are removed and replaced by culture medium which decreases the cell concentration. To account for this change and analyze such types of experiments, we introduced an additional term in each of the Equation 6.1: $-k_S C$ and $-k_S I$. Since the maintenance of healthy tissue cultures is meant to keep the cell concentration approximately constant at least in the initial stages of infection, the splitting constant, k_S, equals the growth rate constant, g. In this case the oscillations disappear as can be seen in Figure 6.4. This may explain why oscillations in tissue cultures of suspension cells which are regularly split are rarely seen.

6.4.2 Spreading Infection by Cell-Free Virions

In this case the following equations were used for numerical simulations:

$$C' = -kv * V * C + g * C * \left[1 - (C - I)/Cm\right] \tag{6.2a}$$

$$I' = kv * V * C - m * I \tag{6.2b}$$

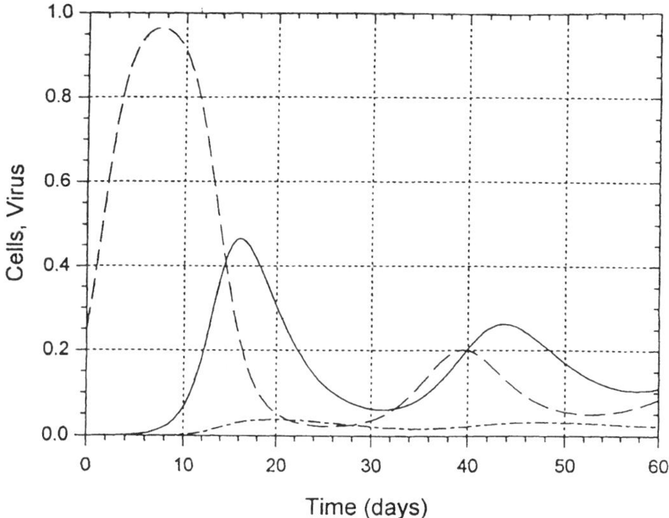

FIGURE 6.5 HIV-1 infection kinetics in tissue cultures by cell-free transmission of the virus as simulated by the solution of Equations 6.2 with parameters: $kv = 10^{-8}$/ml/day, $n = 2 \times 10^3$/day, $ka = 2 \times 10^{-5}$/ml/day, and $mv = 1$/day. The dimensionless variables, the concentration of infected (solid lines) (I/Cm, initial value = 0.25), uninfected (dashed lines) (C/Cm, initial value = 0) cells, and of virions (short dashed-dotted line) (V/Vm, $Vm = 2 \times 10^9$/ml, initial value = 0.00025), are shown as functions of the dimensional independent variable time (days). (From Spouge, J. L. et al., *Bull. Math. Biol.*, 138, 1, 1996.)

$$V' = n * I - ka * V * (C + I) - mv * V \tag{6.2c}$$

where kv (10^{-8}/ml/day) is a rate constant for cell-free infection, V = virion concentration, n (2×10^3/day) = virion production rate constant, ka (2×10^{-5}/ml/day) = attachment rate constant and mv (1/day) = virus inactivation rate constant. The numerical run of Equations 6.2 (Figure 6.5) shows that the number of infected cells changes in a similar fashion as for the case of cell-to-cell spread. Interestingly, the virus load is proportional to the number of infected cells, which means that a quasi steady state establishes relatively quickly. One might note that accurate determination of kv requires specifically designed experiments and depends strongly on culture conditions. In this case, the value 10^{-8}/ml/day has been chosen for comparative purposes assuming that the kinetics of cell-free infection is about the same as for cell-to-cell transmission.

6.4.3 Implications for Understanding HIV Infection Kinetics

The model of HIV-1 infection kinetics, presented above,[23] describes quantitatively or semiquantitatively characteristic features of HIV-1 spread in tissues cultures. While more experimental work in strictly defined conditions is needed to resolve some issues, it appears that our approach based on phenomenological description of virus cell interactions represents semiquantitatively the experimentally observed pattern of virus spread. The model predicts that in a typical tissue culture infection, which starts with a relatively small number of infected cells, the initial growth of virus is exponential and after reaching a peak the viral load either decays to zero or enters an oscillatory pattern. Due to differences in the rates of elimination of virus and cells, a quasi steady state establishes where the concentration of virus in the culture supernatant is proportional to the number of infected cells. On this general background several important considerations must be pointed out.

First, the cell-to-cell spread of HIV-1 in tissue cultures is likely the dominant mechanism of virus spread in most experimental designs.[24,29,30] However, in a typical cell culture experiment the cells tend to cluster and do not make "free diffusion" contacts. This effect is particularly well demonstrated when the cell concentration is decreased. Even cell concentrations as low as 100/ml may lead to cell aggregation.[24] Inside the cell aggregate the cell concentration is high and the virus spread is very efficient. However, outside the aggregate the cell concentration is practically zero and the transmission of HIV-1 can occur only by cell-free virions, which is not so efficient as cell-to-cell transfer. The cell aggregation effect is partially overcome when the cells are split, typically every 2nd day. In many cases the aggregates are disrupted and the cells are well mixed during splitting. Therefore, the typical cell culture system behaves as a partially mixed one, and in some cases more complex equations should be used to describe the transmission of HIV-1 in a heterogeneous system.

Second, the prediction of oscillations in the virus and cell concentrations is consistent with previous models of HIV-1 infection *in vivo*.[16] It has been concluded, however, that reasonable values of the parameters exclude the occurrence of oscillations in infected individuals.[16] Our results demonstrate that this is not the case with tissue culture infections. Our model predicts oscillations at parameters which are quite reasonable for tissue culture experiments. The basic mechanism of the oscillations is the predator–prey type of interactions between the infected cells (or virus) and uninfected cells. With an increase in the number of infected cells, the concentration of the uninfected cells decreases. This leads to a decrease in the rate of production of infected cells. With fewer infected cells, it becomes less likely for the uninfected cells to be infected and they begin to increase in number. This leads to a higher probability of being infected and to the subsequent increase of the infected cells, which completes the oscillation cycle. The frequency of the oscillation depends on the rate of growth and rate of death, but, interestingly, not on the rate of infection because the rate of increase in the number of infected cells equals the decrease of uninfected cells due to infection, and the two effects cancel each other.

While such oscillations are frequently observed in experiments with tissue cultures and the predator–prey explanation is likely to hold, other effects may also lead to oscillations. One possibility is the existence of defective interfering virions. In the case of HIV-1, this effect is particularly important because of the high ratio of defective to infecting virions (more than 1000 to 1[24,32]). In this case, the release of a large amount of defective virions or their products, e.g., free gp120, may lead to their binding to the receptor molecules (e.g., CD4) on the surface of uninfected cells and protect them from subsequent infection. This will lead to a decrease in the number of infected cells and the corresponding decrease of the virus load, which in turn will decrease the protection of the uninfected cells and more infection of healthy cells. This will increase the virus load and the cycle will repeat. Our numerical estimations show that this effect is possible at times close to the peak of infection when most of the cells are infected; at this time the concentration of gp120 could be about 1 μg/ml which is sufficient to protect CD4+ cells from infection. Other effects include existence of cell subpopulations, viral strains with different infectivity, etc. Future work should address these effects if they are significant.

Third, while the resemblance of the cell culture infections to the *in vivo* infections is striking with respect to some parameters, e.g., death rate of acutely infected cells, the general pattern of the HIV-1 infection *in vivo* may be entirely different. In addition to the obvious lack of immune system in tissue cultures and a number of regulatory pathways, the *in vivo* infection occurs in a multicompartmental system where the exchange of cells and virus between the different compartments follows rules which are not completely understood. Therefore, any conclusion made for the tissue culture infections should be very carefully considered before applying it to the more complex infections of whole organisms.

Measurements of changes in viral load and CD4 cells after application of potent HIV-1 inhibitors in patients with AIDS[33,34] found that the half-life of cells productively infected with HIV-1 is about 2 days; this coincides with the infection cycle time (3 to 4 days) of acutely infected cells in tissue cultures (4/2 = 2).[24] In the next section these results will be discussed in more details.

6.5 HIV INFECTION KINETICS IN HUMANS

6.5.1 Estimation of Kinetic Parameters in Patients with AIDS

Recent studies of HIV-1 infection kinetics in patients with AIDS have provided important experimental data describing changes in plasma viral RNA and numbers of circulating CD4+ T lymphocytes following treatment with potent inhibitors of the viral protease or reverse transcriptase.[21,33,34] It was reported that the levels of cell-free virus in the plasma fell rapidly within the first days and weeks after the start of antiviral therapy, which was accompanied by a rise in the number of CD4+ lymphocytes. The authors assumed that the rise in CD4+ cell number observed equaled the number of cells that were *not* infected and killed by HIV-1 as a consequence of the effective therapy. Based on this assumption, they calculated that HIV-1 infected and killed approximately 4×10^7 CD4+ PBLs each day. It was also proposed that the measured rise in circulating CD4+ cells reflected an increase in the total number of CD4+ lymphocytes in all body compartments of an infected individual. Since only 2% of lymphocytes are present in the peripheral blood, the authors multiplied the CD4+ cell increase by 50 and calculated that a patient with HIV-1 AIDS can replace approximately 2×10^9 CD4+ cells each day.

As was discussed in the previous section, during a single cycle each infected cell generated approximately 10^3–10^5 physical particles and 10 to 100 infectious particles, depending on the particular isolate used. The time required to complete one cycle of infection (t_i) calculated for several independent tissue culture infections (3 to 4 days) is strikingly similar to the reported half-life (1.5 to 2 days) of HIV-1 producing cells in individuals responding to potent antiviral agents (3/2–4/2).[33,34] Because one could legitimately question the validity of applying infection rate constants derived from tissue culture infections for the quantitation of HIV infectivity *in vivo*, we also calculated the *in vivo* infection rate constants and n values. The rate constant for the exponential rise of mutant virus in patients with AIDS was 0.27.[34] We calculated very similar values (0.3/day) of the rate constant for patients with primary HIV-1 infection and for monkeys. By assuming a cycle time of 3 days, the calculated number of IUs $n = \exp(kt_i) = \exp(0.9) = 2.5$. This number is close to the minimal number of IUs ($n = 1$) which still allows spreading infection. Therefore, our analyses revealed that, on the average, two to three infectious HIV-1 particles were released from each cell in the two types of *in vivo* infection. An independent quantitative study, using an end point dilution approach, reported that one infectious particle corresponded to about 6×10^4 physical particles in HIV-1-infected individuals.[26] Thus, in *both* tissue culture and *in vivo* infections, an infected cell can produce approximately 10^3 to 10^5 physical particles and far fewer infectious virions.

The two reports monitoring the reduction of cell-free HIV-1 following the institution of antiretroviral treatment estimated the number of newly infected/replaced $CD4^+$ cells in the peripheral blood of patients with AIDS to be 4×10^7/day.[33,34] As noted above, this number of cells would be capable of generating 10^{11} to 10^{12} ($4 \times 10^7 \times 10^4$) physical virus particles. However, the plasma virus load measured by both groups was only 10^8 to 10^9 particles, each having a half-life of about 0.3 days.[21,35] This amount of virus would be produced by 10^4 to 10^6 not 4×10^7 $CD4^+$ cells/day; at the quasi steady state the number of productively infected cells $N = V^*k/n$, where V = number of virions, n = number of virions produced by one productively infected cell per day, and k = virion clearance rate constant equal to $\ln 2/t_{1/2}$, $t_{1/2}$ being the half-life of the virus. Alternatively, the half-life of the virus should be on the order of seconds or less, which is unrealistic. Unfortunately, neither group directly measured the number of productively infected cells in their analyses of virus and $CD4^+$ lymphocyte turnover during antiviral therapy. The number of newly infected cells per day was estimated from increases in $CD4^+$ cell number, assuming, as noted above, a quasi steady state relationship between $CD4^+$ lymphocyte replacement and killing. Recent data showed that the number of productively infected cells in the whole body could be in the range of 10^7 to 10^8, predominantly located in the lymphoid tissue.[36] If one assumes that 2% of these cells are in the blood, then the number of productively infected cells in the blood is about 10^5 to 10^6, which means that about 1 out of 10^3 to 10^4 CD4 cells in patients with AIDS (with less than 200 CD4 cells/μl) is productively infected; this number is consistent with early reports measuring directly the number of productively infected cells[37,38] and with the estimate presented above based on measured viral load and virus half-life.

6.5.2 Analysis of CD4 Cell Regeneration Rate

While the measured rise in circulating CD4+ cells after antiviral drug treatment in many cases is substantial, this increase may not be representative of total CD4+ cell turnover. Lymphocyte trafficking, homing, and recirculation is a complex, multifactorial process and changes in the CD4 cell number in the peripheral blood may not reflect simple quasi steady state alterations. Because circulating lymphocytes comprise such a small fraction (1/50) of the total lymphocytes in the body, major changes can occur (e.g., an increase of 500 CD4+ cells/μl over a 10-h period[39]) as a consequence of the rapid redistribution of cells from different compartments that attends normal diurnal variation.[40] In general, a multifactorial system, such as the lymphoid system, containing lymphocytes that have the capacity to circulate throughout the body or to reside, for extended periods of time,[41] in spatially heterogeneous environments, responds to disturbances of its quasi steady state in a complex manner, best described by exponential functions with rate constants reflecting activity in different compartments. Thus, the 50-fold multiplication step to calculate the total number of CD4+ cells killed by HIV-1 and replaced each day assumes rapid and complete exchange of lymphocytes between the peripheral blood and the rest of the body. This assumption is not warranted[41] for such a large, heterogeneous, nonequilibrated system unless the numbers and dynamic properties of cells residing in diverse major compartments are directly measured.

Our recent analysis of the increase in CD4 cells after intensive chemotherapy seems to support the notion that CD4 increases after treatment with antiviral agents are due to redistribution rather than regeneration.[42] The doubling times of CD4 cells regeneration were in the range from 20 to longer than 200 days. These numbers are larger than the doubling times (about 15 days) calculated from the increases in CD4 cell numbers after treatment with antiviral drugs. While it is possible that patients may show reduced CD4 cell regenerative capacity after intensive chemotherapy due to marrow and/or thymic damage, such differences could also occur if changes in CD4 cells after antiviral treatment were related to altered lymphocyte trafficking. Trafficking dynamics is expected to change more rapidly with time, allowing fast adaptation to changing environments, than regeneration from progenitor populations.

6.5.3 A Model of HIV Concentration Changes Due to Diurnal or Other Physiological Variations

We also developed a model of HIV concentration changes due to physiological causes, including diurnal variations.[35] The model, which is described below, predicts variations in viral RNA concentrations and provides estimates of virus half-life when compared with data, without the use of antiviral agents.

After the primary infection, a quasi steady state is established where the production of virions is about the same as its clearance.[43] This quasi steady state can be perturbed, and the response of the physiological system to the perturbances can provide information for parameters important for infection kinetics and pathogenesis. The steady state can be disturbed in a variety of different ways. Perturbations by antiviral agents resulted in an exponential decrease in the amount of virus and provided an estimate for the virus half-life.[21,33,34] In the model described here we consider perturbations of the steady state not due to drugs, but instead due to natural variations. The analysis is valid for variations, both periodic and nonperiodic, produced by any mechanism which causes a change in virus concentration, for example, changes in virus production after hormone treatment or immune stimulation, probably due to changes in transcription, or diurnal variations in the number of T4 cells, leading to a change of the virus concentration in the blood. To explore these possibilities we developed a mathematical model based on the following assumptions:

1. The rate of virus production is proportional to the number of infected cells I. The proportionality coefficient, k, can be a function of external stimuli which could depend on time.
2. The clearance rate is proportional to the virus concentration V.
3. The number of infected cells, I, changes.

The changes in the number of infected cells could follow any functional dependence following any stimulus that produces a transient increase in viral RNA concentration. The viral RNA concentration could then decay back to its baseline level. This decay can be described using a first-order kinetic model similar to that used to describe the viral RNA decrease following antiretroviral therapy.

These assumptions (for simplicity, we assume that the concentration of infected cells is represented by a periodic sine function) lead to the following equation, describing the variation in the number of virions:

$$dV/dt = kI - cV \tag{6.3}$$

where t = time, k = virus production rate constant, c = clearance rate constant, and

$$I = I_a + A\sin(ft + a) \tag{6.4}$$

Here, I_a is the average number of infected cells in the quasi steady state, A is the amplitude of its variation, f is the frequency of variation (equal to 0.262/h for diurnal variations) and a is a phase constant. The variation represented by Equation 6.4 resembles that experimentally observed for T4 cells,[39] where the maximal CD4 cell number occurs at about 16:00 for patients infected with HIV-1. Here we define $t = 0$ at midnight and, therefore, for this particular case the phase constant $a = -2.59$. The substitution of Equation 6.4 into 6.3 and solving the resulting differential equation for its quasi steady state leads to

$$V = V_a + \left(Akc\cos b/\left(f^2 + c^2\right)\right)\sin(ft + a - b) \tag{6.5}$$

where V_a is the average virus concentration for a single day at the quasi steady state equal to

$$V_a = kI_a/c \tag{6.6}$$

and tg $b = f/c$, b being a phase constant due to the finite rate of virus clearance. The existence of a phase constant means that the changes in virus concentration will follow the changes in the number of infected cells with some delay equal to b/f.

The relative amplitude of the viral RNA variation $(V - V_a)/V_a$ can be then calculated by combining Equation 6.5 and 6.6 as

$$V_{\mathrm{rel}} = \left(V - V_a\right)/V_a = \left(1/\left(1 + f^2/c^2\right)\right)\left(A/I_a\right)\cos b \, \sin\left(ft + a - b\right) \tag{6.7}$$

It is seen from Equation 6.7 that V_{rel} critically depends on the ratio of the frequency of the periodic variations, f, and the clearance rate constant, c. When the clearance rate constant is smaller (which corresponds to long virus half-life) than the variation frequency, i.e., f/c is large, the relative amplitude of virus number variation is small. This is due to both terms, $1/(1 + f^2/c^2)$ and $\cos b$, which decrease with an increase in f/c. However, when f/c is on the order of 1 or smaller, these terms increase and become on the order of 1. In the limiting case of very high clearance rates, the relative amplitude in the viral RNA variation equals that of the relative amplitude of variations in the number of infected cells A/I_a

$$V_{\mathrm{rel}} = \left(V - V_a\right)/V_a = \left(A/I_a\right)\sin\left(ft + a\right) \tag{6.8}$$

In this case, there is no delay in the response and the variation in the number of virions follows the same time dependence as the variation in the number of infected cells. However, when the virus clearance rate c is comparable to the frequency f, i.e., $c = f$, then tg $b = 1$ and $b = 0.78$. This means that the response in virus change will be delayed by $t = b/f = 0.78/0.26 = 3$ h. For this particular case, $b = 0.78$, the maximal virus concentration will appear at 19:00 if the maximal number of infected cells is at 16:00. Figure 6.6 shows an example of relative changes in the virus concentration $(V - V_a)/V_a$ as a response to the relative changes in the concentration of infected cells $(I - I_a)/I_a$ for $b = 0.78$, $f = 0.26$, $a = -2.59$, and $A/I_a = 0.3$ as calculated from Equations 6.7 and 6.4.

The mathematical model predicts oscillations in the virus concentration if the source of virus oscillates. The relative amplitude of these variations is comparable to the relative amplitude of the source of virus if the clearance rate is on the order of magnitude of the frequency of diurnal variations or higher. The clearance rate constant can be estimated by either measuring the relative amplitude or the delay in the virus concentration change.

It has been previously observed that the number of CD4 cells in adults varies throughout the day; there is a peak of CD4 cells in the evenings and a minimum in the mornings, although in adult patients with HIV-1 this pattern is somewhat blunted.[39,44-46] Diurnal variations in the number of productively infected cells have not been reported probably as a result of the technical difficulties of accurately measuring small numbers. However, one may expect that the number of those cells also will fluctuate even if they lose the CD4 marker, because the diurnal variations do not specifically depend on expression of CD4.[47] If one assumes that the peak in

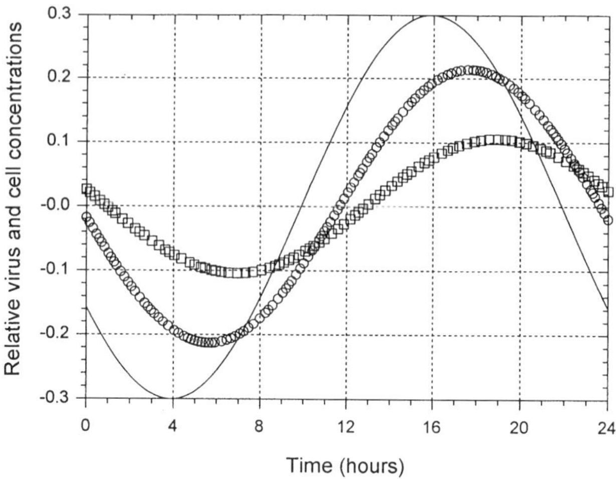

FIGURE 6.6 Variations of viral RNA (○, $c = 0.262$; □, $c = 0.524$) calculated as functions of time in response to variations in the number of infected cells or other sources of variation (continuous line). The parameters values used to calculate the curves from Equations 6.4 and 6.7 are $A/I_a = 0.3, f = 0.262/\text{h}, a = -2.59$. (From Zeichner, S. L. et al., *Pathobiology*, 351, 1997.)

the number of infected cells is about 16:00, then as shown above the mathematical model predicts that the maximum in the viral load can occur 3 h later (at 19:00) only if $f = c$, i.e., the half-life of the virus $t_{1/2} = 0.69/c = 0.69/0.26 = 2.7$ h. In addition, the maximal amplitude of the relative virus concentration variations $(1/(1 + f^2/c^2))(A/I_a)\cos b$, see Equation 6.7) will be much smaller than the maximal amplitude of the relative variation in the number of infected cells A/I_a if $f/c \gg 1$. Because our experimental results indicate that the amplitude of the virus concentration variation is in the range of 0.1 to 0.6[35] and the amplitude of the variation in the number of CD4 cells does not exceed 0.4 to 0.5,[44] f is either much smaller than or comparable to c. This again leads to the conclusion that the half-life is on the order of several hours or shorter. This analysis can be done in the reverse order. On the basis on the experimental results we can conclude that whatever the source of variation in the virus concentration was it has a maximal amplitude several hours before the maximum in viral RNA. Undoubtedly, this kind of analysis will benefit tremendously from future experiments aimed at measuring the number of infected cells and other parameters of the system.

The magnitude of the variations is large enough to develop estimates of the kinetic parameters, but it is not large enough to affect significantly the estimates of kinetic parameters obtained through the use of antiviral agents, where the viral RNA concentration falls by more than two orders of magnitude over a few days.

Recently, we provided evidence that diurnal variations in viral RNA exist in pediatric patients with AIDS.[35] If the hypothesis that those diurnal variations are due to fluctuations in the number of CD4 cells in the blood is correct, one can argue that (1) most of the virus in the blood is produced by productively infected cells in the blood (i.e., the large amount of virus produced in tissues[48] may not make a large contribution to circulating virus) and (2) the virus clearance rate cannot be much lower than the diurnal frequency (equal to 0.26), i.e., the half-life of the virus is of the order of or shorter than several hours. Interestingly, this estimate of clearance rates, obtained by the observation of natural variations in virus level, agrees well with the range in clearance rate (corresponding to half-lives shorter than 2 days) obtained through the use of potent antiviral agents.[21,33,34]

These results may have important implications for understanding mechanisms of pathogenesis. For example, one might speculate that if the immune system is capable of clearing only a limited amount of virus, then fluctuations in virus levels may permit a temporary escape from those clearance mechanisms.

6.6 IMPLICATIONS FOR TREATMENT AND VACCINE DEVELOPMENT

While mathematical modeling of HIV infection kinetics, unlike molecular modeling, may not have a direct impact on development of therapeutics and candidate vaccines, it certainly contributes to our understanding of how HIV spreads and may cause disease. The better understanding of virus infection kinetics may help in the design of treatment protocols and vaccine development. It appears, however, that again, as in the early days of "mathematical" virology, "simple" estimates, but not sophisticated models, have major impacts on our appreciation of HIV dynamics. Thus, the findings that the half-life of the majority of HIV-producing cells is relatively short (about 2 days)[33,34] and that the virus half-life is even shorter (about several hours)[21] lead to the concept that HIV undergoes many rounds of replication and thus may generate a large number of mutants.[43] Therefore, to avoid drug resistance it has been proposed that treatment must start as early as possible with the most-effective drug combinations in spite of the possibility for higher toxicity of drugs in combinations. For the same reason, the immune response following vaccine administration must be as vigorous and rapid as possible from the very beginning of the infection. Because the CTL response is very rapid and in general more efficient against viral infections, it appears that potent vaccines must be developed which induce cellular immunity.

At present, two major problems in fighting AIDS are emergence of drug resistance during treatment with antiviral drugs and lack of efficient vaccines able to protect against a variety of viral isolates. A major challenge for mathematical modelers is to help in solving these problems. One such a problem is how many drugs could prevent emergence of drug resistance. Estimates based on mutation rates and number of infected cells indicate that three drugs (if they completely block HIV-1 infection) may prevent drug resistance in the majority of patients (D. Dimitrov, unpublished calculations). Another related question is how the existing drugs must be applied (temporal order and concentrations) to ensure the less likely conditions for appearance of drug resistance. Mathematical modeling can certainly help in answering these and other questions.

ACKNOWLEDGMENT

I would like to thank my many colleagues and collaborators for sharing their ideas. I am especially grateful to my long-time collaborators R. Blumenthal, M. Martin, J. Spouge, G. Englund, R. Wiley, and S. Zeichner for exciting discussions, as well as to S. Merril for suggestions about the stochastic nature of viral infections. I appreciate the help of my daughter, Dimana, who edited the manuscript and drew Figures 6.1 and 6.2, and my wife, Mariana, for comments.

REFERENCES

1. Delbruck, M., Adsorption of bacteriophage under various physiological conditions of the host, *J. Gen. Physiol.,* 23, 631, 1940.
2. Delbruck, M., The growth of bacteriophage and lysis of the host, *J. Gen. Physiol.,* 23, 631, 1940.
3. Dulbecco, R., Production of plaques in monolayer tissue cultures by single particles of an animal virus, *Proc. Natl. Acad. Sci. U.S.A.,* 38, 747, 1952.
4. Ellis, E. L. and Delbruck, M., The growth of bacteriophage, *J. Gen. Physiol.,* 22, 365, 1939.
5. Valentine, R. C. and Allison, A. C., Virus particle adsorption I. Theory of adsorption and experiments on the attachment of particles to non-biological surfaces, *Biochim. Biophys. Acta,* 34, 10, 1959.
6. Dimitrov, D. S., Dynamic interactions between approaching surfaces of biological interest, *Prog. Surf. Sci.,* 14, 295, 1983.
7. Levy, J. A., *HIV and the Pathogenesis of AIDS,* ASM Press, Washington, D.C., 1994.
8. Coffin, J. M., Ed., *Retroviruses,* Cold Spring Harbor Laboratory Press, Plainview, NY, 1997.
9. Cooper, L. N., Theory of an immune system retrovirus, *Proc. Natl. Acad. Sci. U.S.A.,* 83, 9159, 1986.
10. Merril, S., AIDS: background and the dynamics of the decline of immunocompetence, in *Theoretical Immunology,* Perelson, A. S., Ed., Addison-Wesley, Redwood City, CA, 1987, 59.
11. Reibnegger, G., Fuchs, D., Hausen, A., Werner, E. R., Dierich, M. P., and Wachter, H., Theoretical implications of cellular immune reactions against helper lymphocytes infected by an immune system retrovirus, *Proc. Natl. Acad. Sci. U.S.A.,* 84, 7270, 1987.
12. Anderson, R. M. and May, R. M., Complex dynamical behavior in the interaction between HIV and the immune system, in *Cell to Cell Signalling: From Experiments to Theoretical Models,* Goldbeter, A., Ed., Academic Press, New York, 1989, 335.
13. McLean, A. R., A model of human immunodeficiency virus infection in T helper cell clones, *J. Theor. Biol.,* 147, 177, 1990.
14. Hraba, T., Dolezal, J., and Celikovsky, S., Model-based analysis of CD4+ lymphocyte dynamics in HIV infected individuals, *Immunobiology,* 181, 108, 1990.
15. Nowak, M. A., Anderson, R. M., McLean, A. R., Wolfs, T. F. W., Goudsmit, J., and May, R. M., Antigenic diversity thresholds and the development of AIDS, *Science,* 254, 963, 1991.
16. Perelson, A. S., Kirschner, D. E., and Boer, R. D., Dynamics of HIV infection of CD4+ T cells, *Math. Biosci.,* 114, 81, 1993.

17. Essunger, P. and Perelson, A. S., Modeling HIV infection of CD4+ T-cell subpopulations, *J. Theor. Biol.,* 170, 367, 1994.
18. Pennypacker, C., Perelson, A. S., Nys, N., Nelson, G., and Sessler, D. I., Localized or systemic *in vivo* heat inactivation of human immunodeficiency virus (HIV): a mathematical analysis, *J. Acquired Immune Deficiency Syndrome,* 8, 321, 1995.
19. Nelson, G. W. and Perelson, A. S., Modeling defective interfering virus therapy for AIDS: conditions for DIV survival, *Math. Biosci.,* 125, 127, 1995.
20. Nowak, M. A., May, R. M., Phillips, R. E., Rowland-Jones, S., Lalloo, D. G., McAdam, S., Klenerman, P., Koppe, B., Sigmund, K., Bangham, C. R. M., et al., Antigenic oscillations and shifting immunodominance in HIV-1 infections, *Nature* (London), 375, 606, 1995.
21. Perelson, A. S., Neumann, A. U., Markowitz, M., Leonard, J. M., and Ho, D. D., HIV-1 dynamics *in vivo*: virion clearance rate, infected cell life-span, and viral generation time, *Science,* 271, 1582, 1996.
22. Phillips, A. N., Reduction of HIV concentration during acute infection: independence from a specific immune response, *Science,* 271, 497, 1996.
23. Spouge, J. L., Shrager, R. I., and Dimitrov, D. S., HIV-1 infection kinetics in tissue cultures, *Bull. Math. Biol.,* 138, 1, 1996.
24. Dimitrov, D. S., Willey, R. L., Sato, H., Chang, L.-J., Blumenthal, R., and Martin, M. A., Quantitation of HIV-1 infection kinetics, *J. Virol.* 67, 2182, 1993.
25. Daar, E. S., Moudgil, T., Meyer, R. D., and Ho, D. D., Transient high levels of viremia in patients with primary human immunodeficiency virus type 1 infection, *N. Engl. J. Med.,* 324, 961, 1991.
26. Piatak, M., Jr., Luk, K. C., Williams, B., and Lifson, J. D., Quantitative competitive polymerase chain reaction for accurate quantitation of HIV DNA and RNA species, *Biotechniques,* 14, 70, 1993.
27. Dimitrov, D. S. and Martin, M. A., HIV results in the frame. CD4+ cell turnover, *Nature* (London), 375, 194, 1995.
28. Haynes, B. F., Pantaleo, G., and Fauci, A. S., Toward an understanding of the correlates of protective immunity to HIV infection, *Science,* 271, 324, 1996.
29. Sato, H., Ornstein, J., Dimitrov, D. S., and Martin, M., Cell-to-cell spread of HIV-1 occurs within minutes and may not involve the participation of virus particles, *Virology,* 186, 712, 1992.
30. Philips, D. M., The role of cell-to-cell transmission in HIV infection, *AIDS,* 8, 719, 1994.
31. Murray, J. D., *Mathematical Biology,* Springer-Verlag, New York, 1989.
32. Layne, S. P., Merges, M. J., Dembo, M., Spouge, J. L., Conley, S. R., Moore, J. P., Raina, J. L., Renz, H., Gelderblom, H. R., and Nara, P. L., Factors underlying spontaneous inactivation and susceptibility to neutralization of human immunodeficiency virus, *Virology,* 189, 695, 1992.
33. Ho, D. D., Neumann, A. U., Perelson, A. S., Chen, W., Leonard, J. M., and Markowitz, M., Rapid turnover of plasma virions and CD4 lymphocytes in HIV-1 infection, *Nature* (London), 373, 123, 1995.
34. Wei, X., Ghosh, S. K., Taylor, M. E., Johnson, V. A., Emini, E. A., Deutsch, P., Lifson, J. D., Bonhoeffer, S., Nowak, M. A., Hahn, B. H., et al., Viral dynamics in human immunodeficiency virus type 1 infection, *Nature* (London), 373, 117, 1995.
35. Zeichner, S. L., Mueller, B. U., Pizzo, P. A., and Dimitrov, D. S., Kinetics of HIV-1 RNA concentration changes in pediatric patients, *Pathobiology,* 64, 289, 1996.
36. Haase, A. T., Henry, K., Zupancic, M., Sedgewick, G., Faust, R. A., Melroe, H., Cavert, W., Gebhard, K., Staskus, K., Zhang, Z., et al., Quantitative image analysis of HIV-1 infection in lymphoid tissue, *Science,* 274, 985, 1996.

37. Schnittman, S. M., Psallidopoulos, M. C., Lane, H. C., Thompson, L., Baseler, M., Massari, F., Fox, C. H., Salzman, N. P., and Fauci, A., The reservoir for HIV-1 in human peripheral blood is a T cell that maintains expression of CD4, *Science,* 245, 305, 1989.
38. Schnittman, S. M., Greenhouse, J. J., Psallidopoulos, M. C., Baseler, M., Salzman, N. P., Fauci, A. S., and Lane, H. C., Increasing viral burden in CD4+ T cells from patients with human immunodeficiency virus (HIV) infection reflects rapidly progressive immunosuppression and clinical disease, *Ann. Intern. Med.,* 113, 438, 1990.
39. Malone, J. L., Simms, T. E., Gray, G. C., Wagner, K. F., Burge, J. R., and Burke, D. S., Sources of variability in repeated T-helper lymphocyte counts from human immunodeficiency virus type 1-infected patients: total lymphocyte count fluctuations and diurnal cycle are important, *J. Acquired Immune Deficiency Syndr.,* 3, 144, 1990.
40. Olszewski, W. L., *In Vivo Migration of Immune Cells,* CRC Press, Boca Raton, FL, 1987.
41. Young, A. J., Hay, J. B., and Mackay, C. R., Lymphocyte recirculation and life span *in vivo, Curr. Top. Microbiol. Immunol.,* 184, 161, 1993.
42. Mackall, C. L., Fleisher, T. A., Brown, M. R., Andrich, M. P., Chen, C. C., Feuerstein, I. M., Magrath, I. T., Wexler, L. H., Dimitrov, D. S., and Gress, R. E., Distinct regeneration pathways for CD4 and CD8 cells after intensive chemotherapy of pediatric patients, *Blood,* 89, 3700, 1997.
43. Coffin, J. M., HIV population dynamics *in vivo*: implications for genetic variation, pathogenesis, and therapy, *Science,* 267, 483, 1995.
44. Mientjes, G. H., van Ameijden, E. J., Roos, M. T., de Leeuw, N. A., van den Hoek, J. A., Coutinho, R. A., and Miedema, F. F., Large diurnal variation in CD4 cell count and T-cell function among drug users: implications for clinical practice and epidemiological studies, *AIDS,* 6, 1269, 1992.
45. Mientjes, G. H., van Ameijden, E. J., van den Hoek, J. A., Roos, M. T., and Coutinho, R. A., Circadian variation of the CD4 count among drug users, *J. Acquired Immune Deficiency Syndr.,* 7, 205, 1994.
46. Holodniy, M., Mole, L., Winters, M., and Merigan, T. C., Diurnal and short-term stability of HIV virus load as measured by gene amplification, *J. Acquired Immune Deficiency Syndr.,* 7, 363, 1994.
47. Vagnucci, A. H. and Winkelstein, A., Circadian rhythm of lymphocytes and their glucocorticoid receptors in HIV-infected homosexual men, *J. Acquired Immune Deficiency Syndr.,* 6, 1238, 1993.
48. Pantaleo, G., Graziosi, C., Demarest, J. F., Butini, L., Montroni, M., Fox, C. H., Orenstein, J. M., Kotler, D. P., and Fauci, A. S., HIV infection is active and progressive in lymphoid tissue during the clinically latent stage of disease, *Nature* (London), 362, 355, 1993.

7 Mathematical Models of Cell Cycles

Kurt W. Kohn and Dimiter S. Dimitrov

CONTENTS

7.1 CELL CYCLE: BASIC EXPERIMENTAL OBSERVATIONS

The cell cycle is of central importance for life on Earth. Cell division of unicellular and multicellular organisms has been of critical importance for their evolution. For unicellular organisms the control of the cell cycle should sense internal and environmental signals, and cells tend to divide as rapidly as possible under appropriate conditions. Multicellular organisms, however, must have very tight control of the cell cycle so as not to allow the destruction of the integrity of the organism as a whole. One might expect, therefore, that with an increase in the complexity of multicellular organisms, the control of cell cycle would become more sophisticated. However, this does not automatically imply that for complex multicellular organisms the internal control and the cell cycle machinery itself should evolve to a high level of complexity. On the contrary, it is reasonable to assume that the evolution of multicellular organism may have led to improvement in the efficiency of the cell cycle and its internal control, while increasing the complexity of the control by extracellular signals produced by other cells of the organism. In the rest of this section we will discuss basic experimental observations which could be described conceptually or/and quantitatively by mathematical modeling. Because of the very large number of papers published on the cell cycle, we will refer the interested reader to consult recent books[1,2] and review compendia[3,4] for references and more details.

Rapidly dividing human cells have a cell cycle that lasts about 24 h; unicellular organisms such as bacteria and yeasts divide more rapidly — the cell cycle can be completed in 20 to 90 min. The cell cycle is divided into two fundamental parts: interphase, which is the longer one, and mitosis, which lasts about 30 min in human cells and ends with the division of the cell. During the interphase the cell grows, which includes the synthesis of new ribosomes, membranes, mitochondria, endoplasmic reticulum, and most cellular proteins. Chromosome replication is restricted to a specific part of interphase, called S phase. In human cells the S phase lasts about 6 h and is preceded by a gap called G1 (which lasts about 12 h) and is followed by another gap called G2. Cells also can arrest early in G1 for a variety of reasons, including starvation or deprivation of growth factors, in a state called G0.

Several cellular systems have been particularly useful for study of the cell cycle: embryonic cells, including frog and sea urchin eggs, yeasts, including budding yeast (*Saccharomyces cerevisae*) and fission yeast (*Schizosaccharomyces pombe*), and immortalized vertebrate cells, including a variety of continuous cell lines.

Mammalian cell fusion experiments suggested that mitotic cells contain factors that induce mitosis when introduced into cells at the other phases of the cell cycle. Cytoplasm of S phase contains factors that induce DNA replication in G1 nuclei; G1 cells did not enter mitosis until G1 nuclei had finished DNA replication, indicating the existence of feedback controls which can arrest the cell cycle at a checkpoint while the phase is not complete. However, the S phase factors were not able to induce a new round of DNA replication in G2 nuclei, suggesting the existence of blocks in rereplication. Apparently, the block to rereplication must be removed, or the S phase stimulator destroyed, during mitosis to allow the G1 nuclei to replicate.

Early embryonic cells of some egg-laying organisms are controlled by a "simple" biochemical oscillator that periodically drives cells into and out of mitosis independently of the DNA replication cycle. This simple cell cycle machinery has two states: mitosis where the maturation (mitosis) promoting factor (MPF) is active, and interphase, where it is inactive. The activation of MPF is induced by cyclin which is continuously synthesized throughout the cell cycle but degraded at the end of each mitosis by a ubiquitin-dependent process stimulated by active MPF. The destruction of cyclin leads to the inactivation of MPF and to the next interphase. Lower cyclin concentrations slow the cell cycle down. However, increasing cyclin concentration hardly speeds the cycle engine up; a second phase (of posttranslational events) in the interphase of the early embryonic cell is required to activate MPF. The rate of cyclin synthesis probably determines the length of the first period but not that of the second. MPF was initially isolated from frog eggs, but later shown to be present in all mitotic and meiotic cells. It is a heterodimer composed of one cyclin B (B1 or B2) and one p34 (see below) molecule, although in crude extracts MPF activity is associated with larger protein complexes. One of the best substrates for MPF is histone H1; MPF activity is commonly measured as the HI kinase activity.

Genetic analysis of temperature-sensitive cell division cycle yeast mutants allowed identification of a number of genes which play critical roles in the cell cycle. The identification of the Cdc28 protein as absolutely required for DNA replication but not for growth in the G1 phase in budding yeast identified a point of G1/S transition (called START in yeasts) where DNA replication becomes insensitive to loss of Cdc28 function. In fission yeast the homologue of Cdc28 is Cdc2 (often called $p34^{cdc2}$ or p34 because of its molecular weight); all eukaryotes contain functional homologs of $p34^{cdc2}$ with about 65% similarity in amino acid sequence. START is regulated by accumulation of G1 cyclins (Cln1 and 2). Cln1 and 2 reach maximum levels just before START and then rapidly decline. All three cyclins are unstable proteins with a half-life of less than 10 min. The synthesis of Cln1 and 2 mRNA is stimulated by a transcription factor composed of two proteins, Swi4 and Swi6, whose activity depends on phosphorylation, most likely by Cdc28–Cln1/2 complexes. Thus, once G1 cyclins exceed a certain level, they activate some of the Cdc28, leading to an increased G1 cyclin production, which activates more Cdc28, resulting in rapid and irreversible passage through START.

Cdc2 (and Cdc28 in budding yeasts) is also critical for entry into mitosis; its mitosis-inducing activity is inhibited by Wee1 and stimulated by Cdc25 (to avoid confusion note that Cdc25 from budding yeasts is not related to fission yeast Cdc25) in proportion to their concentrations. While Cdc2 and Wee1 are protein kinases, Cdc25 is a protein phosphatase. The concentration of Cdc25 in fission yeasts oscillates in a manner similar to that of cyclin B (in early frog embryos Cdc25 concentration is constant throughout the cycle, although its activity varies).

The p34 subunit of MPF undergoes changes in phosphorylation on tyrosine 15 (Tyr-15), and threonine 161 during the cell cycle; the inhibitory phosphorylation on Tyr-15 is dominant over the stimulatory one on threonine 161. In fission yeasts Cdc25 and Wee1 are major enzymes that regulate phosphorylation on Tyr-15, although even in the absence of Wee1, another kinase, Mik1, can catalyze some tyrosine phosphorylation. The phosphorylation of these residues normally requires association of p34 with cyclin.

In mammalian cells, proliferation begins with the binding of growth factors to their receptors which through a cascade of events, including Ras inactivation, leads to transcription of early response genes which in turn induce the transcription of delayed response genes. The products of the delayed response genes include G1 cyclins and relatives of the yeast p34 that interact with each other to yield active protein kinases which inactivate tumor suppressor gene products and activate the DNA replication machinery resulting in passage through the restriction point. Four classes of putative G1 cyclins in mammalian cells have been identified: cyclins C, D, E, and F. Members of the cyclin D and E family appear to be the best candidates for delayed response genes. Cyclin D genes are a small family of closely related genes, including one oncogene. However, although their transcription can be induced by growth factors, their concentration does not vary throughout the cell cycle. The concentration of cyclin E rises and falls during the cell cycle; an increase in concentration leads to shortening of G1.

Passage through the restriction point in mammalian cells appears to be stimulated by complexes of G1 cyclins and relatives of the yeast p34 called cyclin-dependent kinase (Cdk) proteins. Cdk2 can form complexes with cyclins E and A and is a delayed response protein; its concentration is relatively constant throughout the cell cycle of rapidly dividing cells, but decreases to undetectable levels in starved cells.

The transcription of both early and delayed response genes is regulated by the transcription factor E2F. During G0 and G1 two tumor suppressors, the retinoblastoma protein (Rb) and a relative called p107, form complexes with E2F which abolishes transcription and prevents passage through the restriction point. Phosphorylation of the Rb–E2F complexes induced by Cdk2–cyclin complexes blocks its ability to bind to E2F, which releases free E2F able to induce transcription. Rb is dephosphorylated as cells leave mitosis and phosphorylated again late in G1 at a time roughly corresponding to the restriction point. Thus, the role of G1 cyclins may be autocatalytic: the ability of G1 cyclin–Cdk2 complexes to inactivate suppressor proteins would increase their transcription and the transcription of other delayed response genes, thus leading to a further increase in cyclin transcription and passage through the restriction point. One must note that in mammalian cells many of the regulatory pathways are redundant and other proteins may be involved in the passage through the restriction points. For example, it has been shown recently that transgenic mice lacking Rb are not growth defective. The molecular machinery of the G1-to-S-phase transition is diagrammed in Figure 7.4, and will be further discussed later in this chapter.

One final point which attracted a lot of attention in the past is the probabilistic nature of cell division; even genetically identical synchronized cells show widely variable cycle times. It was discussed that the initiation of the S phase, particularly the first pair of DNA replication forks, could be the all-or-none trigger. Such a single initiation event might depend on a rare random collision between an initiator molecule present at low concentration and a replication origin sequence, thus leading to a random variation of the time spent on the G1 phase.

7.2 CELL CYCLE MODELING: AN OVERVIEW

The importance and the fascinating nature of the oscillations during cell division led to many attempts at mathematical modeling of the cell cycle. While the initial models tried to catch the basic features of oscillations and demonstrate the conceptual possibility for oscillations, recent models are very detailed in nature and could be used for prediction of phenomena previously not observed by experimenters and for design of new experiments. Thus, with respect to the level of complexity of the models, they could be divided into two broad categories: "simple" and "complex." The recent wealth of information about molecular components and interrelations between them now allows development of more-realistic complex models and this trend may well continue in the future. Models can also be divided into several categories with respect to the type of cell cycle they describe. While it now appears that a biochemical oscillator is the driving force of the cell cycle process in all cells, some particular features of this oscillator can differ significantly; e.g., in embryonic eggs the biochemical oscillator is independent of DNA replication, unlike in yeasts and mammalian cells where checkpoints restrict the cycle until certain conditions are met. Finally, another classification of the models is based on the mechanism of oscillations including those based on positive feedback or negative feedback. For more details about those two types of models, development of concepts, and references, see recent reviews by Goldbeter et al.[5,6] In the rest of this section we will briefly overview several examples of models which may represent typical features of the experimental observations as discussed above.

Goldbeter[5,7] developed a model based on the hypothesis that the driving force of the biochemical oscillator is a negative feedback produced by cyclin-activated Cdc2 able to degrade cyclin, coupled to thresholds and time delays. Such systems keep oscillating because threshold levels of both cyclin and Cdc2 trigger reactions with built-in time delays, and the degradation of cyclin induces loss in Cdc2 kinase activity. The model of the cell cycle in early amphibian embryos assumes that cyclin is synthesized at a constant rate and triggers the activation of Cdc2. The formation of the Cdc2–cyclin complex is not explicitly described, but rather it is assumed that Cdc2 is activated by an "activase," e.g., Cdc25, driven by cyclin. The activated Cdc2 is "inactivated" by other molecules, e.g., Wee1, with a constant maximum activity throughout the cycle. It is further assumed that the active Cdc2 activates a cyclin protease by phosphorylation. The protease is inactivated by a phosphatase with a constant maximum activity throughout the cycle.

Thus, the three variables in the Goldbeter model are cyclin, active Cdc2, and active cyclin protease. They are related by three nonlinear differential equations, where all nonlinearities are of the Michaelian type. The analysis of these equations showed that they provide an explanation for the origins of thresholds in the control of Cdc2 by cyclin and the activation of the cyclin protease by Cdc2, and that the thresholds are linked to the time delays which play a primary role in the initiation of oscillations. It describes the cell cycle as follows. Cyclin (starting with a low concentration) initially accumulates almost linearly until it reaches a threshold value at which Cdc2 abruptly becomes activated. The fraction of active Cdc2 increases very rapidly until it reaches another threshold value at which a steep increase in the active cyclin protease occurs. The abrupt activation of the cyclin protease leads to a rapid decline in the cyclin concentration. When the cyclin concentration drops below its threshold value, the active Cdc2 becomes inactivated. Once the active Cdc2 concentration decreases below its threshold value, then the cyclin protease is inactivated. The protease inactivation allows cyclin to begin accumulating again, and the cycle repeats. These oscillations do not depend on the initial conditions and reach the same oscillatory regime. The model demonstrates that the rate of cyclin accumulation largely determines the G1 phase length. It was also shown that while only one threshold can still lead to oscillations, two thresholds (in this case of cyclin and active Cdc2) allow oscillations for a much larger domain of parameter values.

The initial minimal cascade model was further extended to include additional phosphorylation–dephosphorylation cycles, including one based on activation of Cdc25. The extended model predicts more-abrupt oscillations but their essential features remain. This and other models show that the mitotic oscillator can be arrested in a variety of ways by inhibiting any of the enzymes controlling the activation of Cdc2, including the cyclin protease.

Although the minimal cascade model shows that the mitotic oscillator could be based solely on negative feedback, positive feedback may also underlie or contribute to the periodic activation of Cdc2 based on its self-activation; most likely both types of feedbacks are involved ensuring that oscillations occur over a wide range of physiological conditions.

The cell cycle in yeast and somatic cells is more complex than in early embryonic cells. However, in spite of the complexity it appears that the basic features of the oscillator in embryonic cells are retained in the more complex system. Recently, Novak and Tyson extended their previous model of cell cycle in *Xenopus* oocyte extracts[8] to the fission yeast.[9] Interestingly, they showed that all the major features of the fission yeast cell cycle can be derived from a model almost identical to that for *Xenopus* oocyte extracts in spite of the fundamental difference between the physiology of the two cell cycles.

Novak and Tyson's model contains one negative and two positive feedbacks which are at the heart of the mechanism of oscillation: MPF (cyclin–Cdc2 complex) stimulates its own degradation by indirectly activating the ubiquitin pathway that degrades its cyclin subunit, and active MPF stimulates its own production by activating Cdc25 and inhibiting Wee1. According to the model, cyclin monomers are synthesized and degraded, and they associate with Cdc2 monomers. The Cdc2 subunit of the dimer is phosphorylated and dephosphorylated by several kinases and phosphatases, including Cdc25 and Wee1; the dimer form phosphorylated only on Thr-167 (MPF) is active and stimulates mitosis and cell division. The kinase Wee1, which catalyzes the inhibitory phosphorylation of Tyr-15, is inhibited by active MPF and by cell size. The phosphatase Cdc25, which removes the phosphate from Tyr-15, is activated by MPF and inhibited by unreported DNA. The ubiquitin pathway by which cyclin is degraded is stimulated by MPF through an intermediate enzyme; this stimulation can be blocked if certain target proteins are not successfully phosphorylated by MPF during mitosis. The START control is modeled by a set of logical instructions and intended only as a switch for the M-phase control system, but does not include explicit description of the G1/S transition.

This model for yeast is translated into an impressive set of 19 differential equations; several of them including nonlinearities of Michaelian type as in Goldbeter's model. It predicts oscillations and provides explanations for a variety of experimental observations. Even more interestingly, the model makes a number of predictions for several yeast mutants that have not yet been established experimentally.

The complexity of the mammalian cell cycle control and the absence of quantitative data for components of the control machinery limit the development of mathematical models, and they have not reached the level of sophistication of the yeast models. However, in the last few years accumulation of data led to several attempts to develop such models, especially for the control of the G1/S phase transition. Here, we will briefly discuss two recent models describing the cell cycle in mammalian cells, especially the transition form G1 to S phase.

Goldbeter[5] described an extension of his previous minimal cascade model to include the transition through the restriction point. He considers the cell cycle as driven by two self-sustained oscillators, one involving Cdc2, the second involving Cdk2. The mechanism of oscillation for the two oscillators is similar and involves cyclins and corresponding kinases. The proper timing of the peaks in the activity levels of the two kinases would result from the inhibition exerted by each of them on the activation of the other. Numerical simulations of this model showed that peaks in Cdc2 and Ckd2 follow at regular intervals; when Cdc2 reaches its maximal activity, Cdk2 is inactive, and vice versa, as observed experimentally.

An elaborate model, based on a system of three differential equations, specifically attempting to describe the control of the G1 phase in mammalian cells has been recently proposed by Obeyesekere et al.[11] The model is based on Rb phosphorylation by a cyclin E–Cdk2 complex, leading to its inhibition and its subsequent dephosphorylation at the end of the cell cycle. Numerical simulation showed that the model can exhibit periodic concentration changes consistent with those seen in the cell cycle and predicts cell arrest in agreement with experimental observations.

7.3 A MODEL OF THE G1/S PHASE TRANSITION CONTROL

About a quarter of a century ago, Temin[12] demonstrated that chicken cells become independent of external mitogenic signals and thereby committed to DNA replication and mitosis during G1 several hours before entry into the S phase. Later, these findings were extended to other systems and Pardee[13] introduced the term restriction point (R) to define the point in G1 after which cells can proliferate independently of mitogenic signals. At present, a large amount of data is beginning to unravel the molecular components of the machinery that controls the transition from the G1 phase to the S phase passing through the restriction point (see the recent reviews[14,15]).

A possible scenario of the molecular events controlling G1 progression and S phase entry includes two major stages:[14,15] (1) induction of cyclin D synthesis by mitogenic signals and (2) release of E2F from its complex with Rb. Cyclin D associates with Cdk4/6 which leads to phosphorylation of the Rb protein and results in dissociation of active transcription factor E2F from an inactive complex with Rb. E2F can activate target genes, including the cyclin E gene. Cyclin E forms a complex with Cdk2 which may promote further dissociation of E2F from Rb–E2F by a positive feedback which may render cells independent of mitogenic stimulation, thereby contributing to the irreversible progression to the S phase. Passage through the restriction point may represent the increase of the Cdk2–cyclin E-induced kinase activity above a critical level.

We present two types of models of the transition through the restriction point. The first utilizes a system of differential equations and may be considered to be a "macroworld" description;[16] it utilizes a set of differential equations which includes some nonlinear terms. The second utilizes an explicit diagram of molecular interactions and corresponds to a "microworld" description. Both models are based on the following assumptions: (1) an external signal initiates phosphorylation of the Rb in the Rb–E2F complex and leads to the release of E2F, (2) E2F acts as a transcription factor that stimulates the synthesis of cyclin E, and (3) cyclin E binds to and activates Cdk2 which then phosphorylates Rb in the Rb–E2F complex releasing free E2F.

7.3.1 "Macroworld" Model

For simplicity, we assume that the cyclin D–Cdk4/6 concentration (denoted as a) rises and falls in a manner determined by external signals and is modeled as

$$a = a_1 * \exp\left[-0.5 * \left(t/ts - tm/ts\right)^2\right] \quad (7.1)$$

where a_1 determines the maximal level of the external signal, *ts* and *tm* are the deviation and mean time at which the signal is applied, *t* is time, and * denotes multiplication. Further, we will not give the units of the quantities; unless specified, all concentrations are in μM and times in days. We also assume that while free Rb is continuously synthesized and degraded, the majority of Rb molecules are in a high-affinity complex with E2F:

$$R' = s - d * R - k * R * F + k_{\_} * RF \tag{7.2}$$

where prime denotes derivative with respect to time, *R* is Rb, *F* is E2F, *RF* is Rb–E2F complex, *s* is the rate of Rb synthesis, *d* is its degradation rate constant, and *k* and *k_* are the on and off rate constants, respectively.

We further assume that the *RF* complex releases free E2F and phosphorylated Rb following interaction with the cyclin D–Cdk4/6 kinase (denoted by *a*) and the cyclin E–Cdk2 complex (EK2)

$$RF' = k * R * F - k_{\_} * RF - a * RF/(1 + Ka * RF) - b * EK2 * RF/(1 + Kb * RF) \tag{7.3}$$

where the constant *b* accounts for the effect of the cyclin E–Cdk2 activity and ensures the positive feedback mediated by cyclin E. The constants *Ka* and *Kb* are related to the Michaelis constants for the phosphorylation reactions by cyclin D–Cdk4/6 and cyclin E–Cdl2, respectively.

Most of the free E2F (*F*) is “produced” by the phosphorylation-induced dissociation of *RF* as described by the following equation:

$$F' = s_1 - d_1 * F/(1 + Df * F) - RF' \tag{7.4}$$

where s_1 and d_1 represent the E2F rates of synthesis and degradation, *Df* is a Michaelis-related constant for the enzymatic degradation of E2F, and the last term is the amount of *F* released by phosphorylation of the Rb–E2F complex (Equation 7.3 expresses *RF'*).

The amount of free E2F is to a large extent determined by the phosphorylation due to the cyclin E–Cdl2 complex; the formation of that complex is given by the following equation:

$$EK2' = k_1 * E * (1 - EK2/K2t) - k_{-1} * EK2 \tag{7.5}$$

where k_1 and k_{-1} are the on and off rate constants for the complex formation, and *K2t* is the total amount of Cdk2, which for simplicity is assumed constant; the term *K2t* – *EK2* then represents the free Cdk2 available for binding.

We assume that the transcription of the cyclin E gene is activated by E2F, and therefore the cyclin E concentration can be described by the equation:

$$E' = c * F/(1 + f1 * F) - d2 * E/(1 + De * E) - EK2' \tag{7.6}$$

where the constant c reflects the transcriptional activation by E2F; $f1$ is the affinity of E2F binding to the cyclin E promoter and accounts for possible saturation effects at high E2F concentrations; $d2$ is a degradation rate constant; and De is a Michaelis-related constant for the enzymatic degradation of cyclin E.

The system of five differential equations was solved by using the Episode integrator and analytical Jacobian (the program Scientist from MicroMath, Salt Lake City). The initial conditions and parameters, listed in the order they first appear in the above equations, are listed below:

$$R = 0.1;\ RF = 1000;\ F = 1;\ E = 1;\ EK2 = 1 \quad \text{at } t = 0$$

$$a_1 = 10,\ ts = 0.01,\ tm = 0.2,\ s = 1,\ d = 10,\ k = 10^4,\ k\_ = 1,\ Ka = 0,\ b = 100,\ Kb = 1,$$

$$s_1 = 100,\ d_1 = 100,\ Df = 0.1,\ k1 = 1,\ K2t = 10^3,\ k\_1 = 1,\ c = 1,\ f1 = 0,\ d2 = 1,\ De = 0.1$$

The initial conditions were chosen to reflect an established (or close to) steady state in an absence of mitogenic signals and positive feedback. Figure 7.1 shows the E2F and Rb–E2F concentrations as functions of time for four possible situations: (1) when a mitogenic external signal inducing cyclin D–Cdl4/6 activity is applied for a short time (a_1 = 10) in the absence of positive feedback provided by the cyclin E–CDK2 complex ($b = 0$); (2) when both external signal (a_1 = 10) and positive feedback (b = 100) exist; (3) when there is no external signal ($a_1 = 0$), but the positive feedback occurs (b = 100); and (4) in the absence of external signal ($a_1 = 0$) and positive feedback (b = 0).

It is seen that the above system of differential equations represents characteristic features of a trigger system with a positive feedback. In the first case the concentration of E2F rises with the application of the external trigger but declines in the absence of a positive feedback (b = 0). The concentration of E2F, however, begins to increase (after a small decline) and reaches a relatively high level in the presence of a positive feedback (b = 100) (the second case). The small decline and subsequent increase in the E2F concentration is a consequence of the effect of the positive feedback which is turned on after a relatively short delay. In absence of an external trigger (a = 0), the concentration of E2F does not change significantly even in the presence of a positive feedback (b = 100) (case 3). In the steady state (case 4), all concentrations are constant.

While this system may represent well some characteristic features of a trigger system which allows the progression through the restriction point after removing the external mitogenic trigger, further experimental data are needed to evaluate the parameters of the system and the effects of other possible components of the G1/S transition. A variety of other effects and alternative pathways are likely to exist which allow the cell to regulate the transition from G1 to S phase finely.[15]

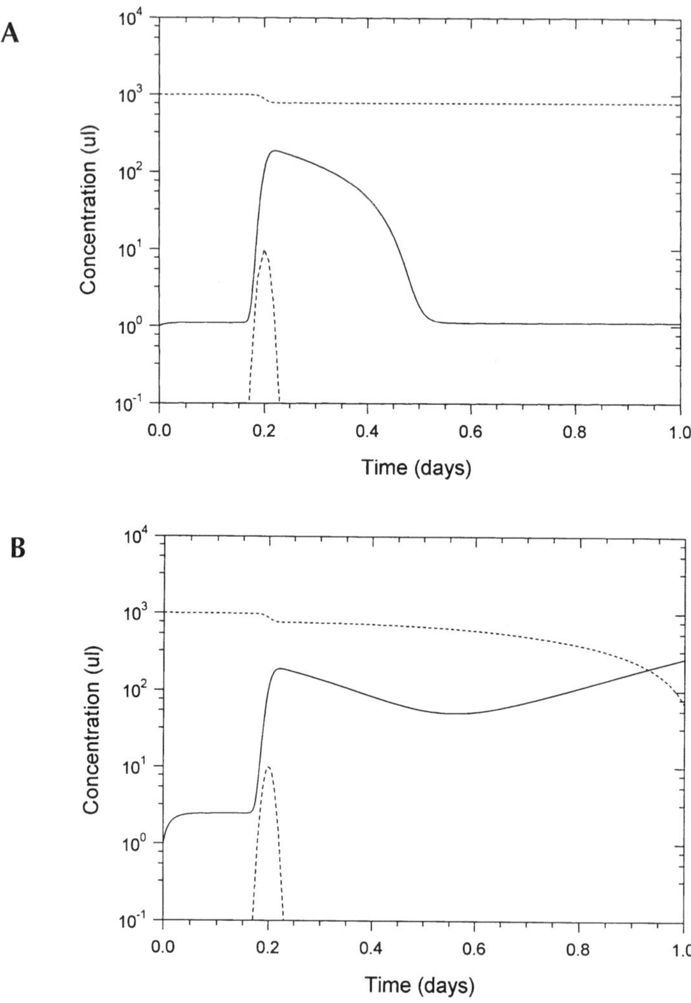

FIGURE 7.1 Computer simulation of the G1-to-S switch in mammalian cells. (A) application of a mitogenic external signal inducing cyclin D–Cdk4/6 activity for a short time (a_1 = 10) in the absence of positive feedback provided by the cyclin E–Cdk2 complex (b = 0), (B) both external signal (a_1 = 10) and positive feedback (b = 100) exist, (C) there is not external signal (a_1 = 0), but the positive feedback occurs (b = 100), and (D) absence of external signal (a_1 = 0) and positive feedback (b = 0). Dots represent Rb-E2F concentration; solid lines are E2F; and dashed lines, the trigger.

7.3.2 "MICROWORLD" MODEL

A convention of reaction diagrams, proposed by one of us (K.W.K.), allows concise and unambiguous description of reaction networks in terms of explicit molecular interactions. These explicit diagrams can generate a digital file that serves as input to a simulation program. A key feature of such microworld descriptions is that the

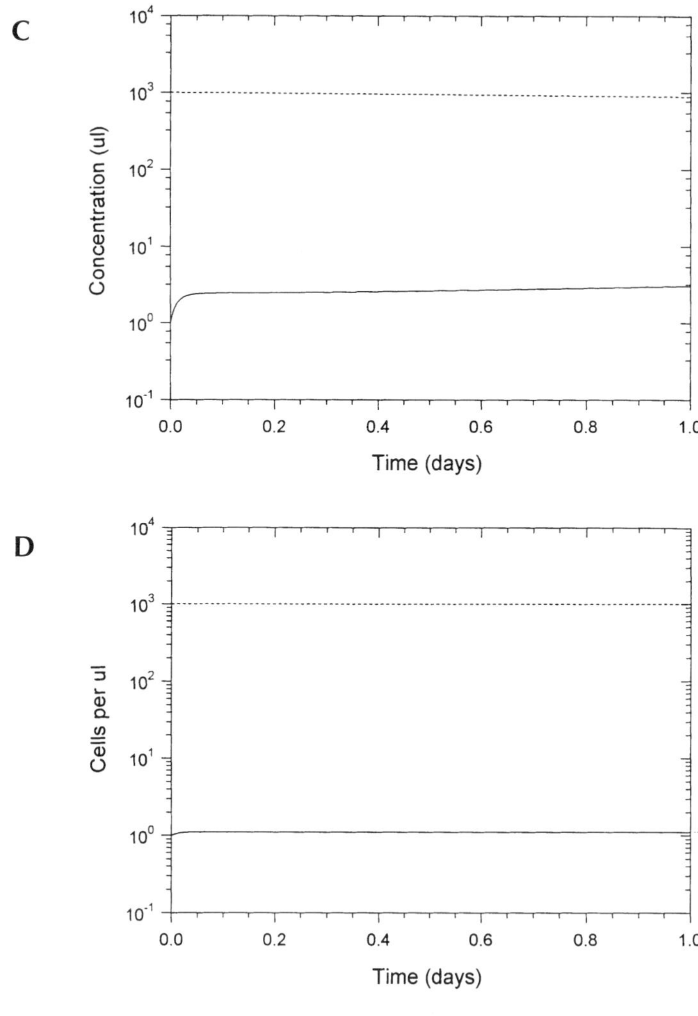

FIGURE 7.1 (continued)

individual reaction steps are simple expressions of the mass action law.[16] Our procedure will now be applied to G1/S phase model discussed above.

An explicit network diagram for the G1/S phase transition, consisting of steps that were implied in the differential equations above, is presented in Figure 7.2. The diagram symbols used are defined in Figure 7.3, and will be discussed more fully later in this chapter in the context of heuristic diagrams.

Initiation of S phase requires the actions of both E2F and cyclin E.[10] The simplest subsystem for production of E2F would consist of the reactions governed by the rate constants k_{18}, k_{19}, and k_{20} (to simplify terminology, we will refer to reactions 18, 19, and 20) (Figure 7.2). Reaction 19 refers to the binding of E2F to a protease (the reverse reaction is assumed not to occur and is therefore designated "0" in Figure 7.2). The degradation of E2F is denoted by reaction 20 which specifies the stoichiometric

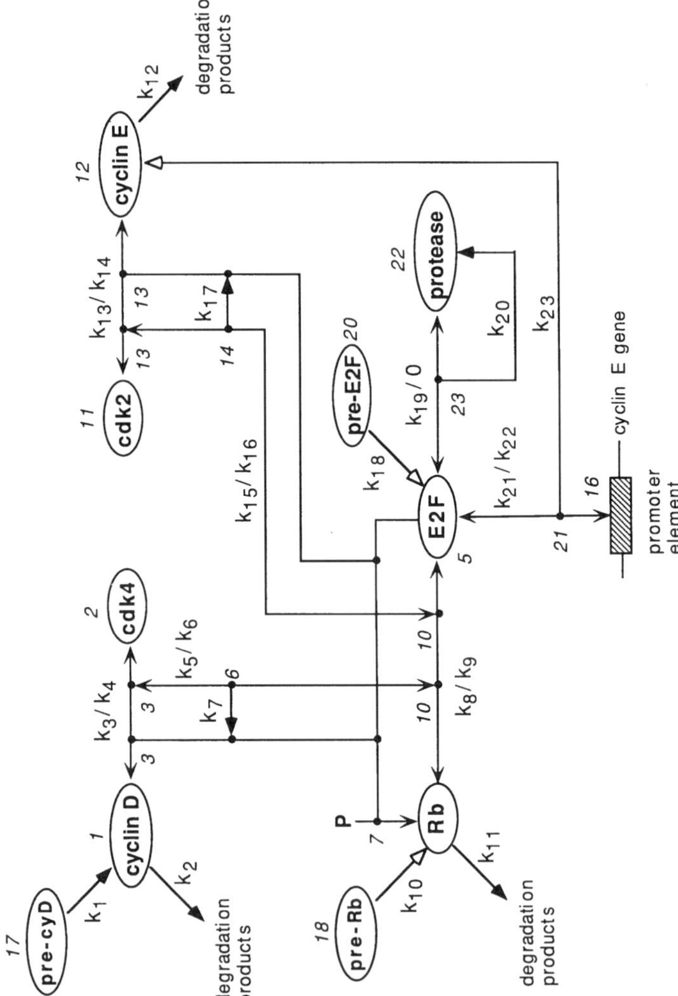

FIGURE 7.2 Explicit network diagram describing a "microworld" model of the G1/S phase transition. The molecular interaction network specified by this scheme is an expanded construct consistent (except in nonessential details) with the preceding differential equations of the "macroworld" description. The symbols used are defined in Figure 7.3.

conversion of E2F–protease complex to protease (E2F being lost in the process). This manner of description corresponds to a Michaelis–Menten term in a differential equation, but avoids steady-state assumptions, and allows all reaction steps to be represented by simple mass–action kinetics. In this model, E2F is assumed to be synthesized at a constant rate; this is achieved by keeping the concentration of "pre-E2F" (species 20) constant (represented by the open arrowhead for reaction 18; the open arrowhead means that product is formed without depletion of "reactant" which thereby assumes instead the role of "stimulator"). At low rates of E2F formation, E2F concentration rises rapidly to a plateau, while at higher rates the concentration of E2F increases further due to the limitation in the degradation rate. The detailed methods and results of simulations will be presented elsewhere.[18]

The next level of development of the E2F regulation subsystem would be the addition of Rb-family protein(s). Rb and its relatives bind to E2F and thereby block E2F activity; phosphorylation of Rb disrupts the binding to E2F and releases active E2F. At the current level of development of the system, however, phosphorylation of Rb is not yet included. If the existence of a pool of Rb (species 4) at the start of a simulation run is assumed, the newly synthesized E2F will bind to available Rb (assuming reaction 8 is much faster than reactions 9 or 19 at early times in the simulation run). Free E2F (species 5) then remains low until free Rb is nearly depleted, whereupon E2F rises rapidly to a plateau. This constitutes a simple model for delayed switching on of E2F.

For the next level of development, we add Rb phosphorylation induced by preformed cyclin E–dependent kinase (cyclin E–Cdk2, species 13). ("Rb" throughout refers to Rb and/or relatives such as p107.) In this simulation, we also include a pool of preexisting Rb. The system is now capable of producing a timed wave of E2F. As E2F is synthesized by reaction 18, it combines rapidly with Rb to form the inactive dimer species 10. The preformed cyclin E–Cdk2 dimer (species 13) combines rapidly with species 10; the resulting tetramer (species 14) then undergoes first-order conversion by reaction 17 to yield three products: phosphorylated Rb (species 7), free E2F (species 5), and regenerated cyclin E–Cdk2. When free Rb is nearly depleted, E2F rises to a plateau whose level is limited by the maximum rate of E2F degradation (reaction 20). By this time, some of the Rb has accumulated as Rb–E2F dimer (species 10) which now decays through dissociation reaction 9 followed by Rb phosphorylation via cyclin E–Cdk2. When the dissociation of Rb–E2F is nearly complete, the E2F degrades and falls to a low level. This simulation exhibits a timed wave of E2F which could represent a primitive G1-S-G2 regulator.

The system can be further embellished by adding the association/dissociation of cyclin E and Cdk2 and the association/dissociation of E2F with E2F-binding promoter element (species 16). The essential behavior of the system is not altered by these additions.

In the next phase of development, we add cyclin E expression from a gene regulated by an E2F promoter element (species 16). We also add cyclin E degradation (reaction 12). In this simulation, we assume that there is no preexisting cyclin E, and that there is a pool of preexisting Cdk2. Thus, a positive feedback loop involving

A

Symbol definitions

Binding (non-covalent, reversible).
A filled circle on the line represents the covalently modified species.

Covalent binding to a site.
A filled circle on the line represents the covalently modified species.

Stoichiometric conversion of one molecular species to another

Stimulation of a process (stimulator not diminished thereby)

Examples:

a. C binds to pre-formed A•B.

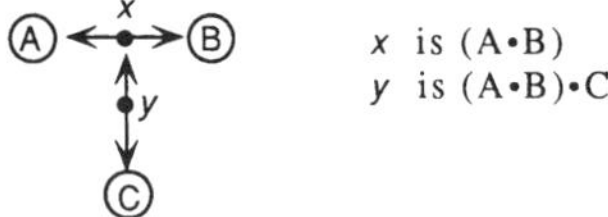

b. Dimers A•B and C•D associate to form a tetramer.

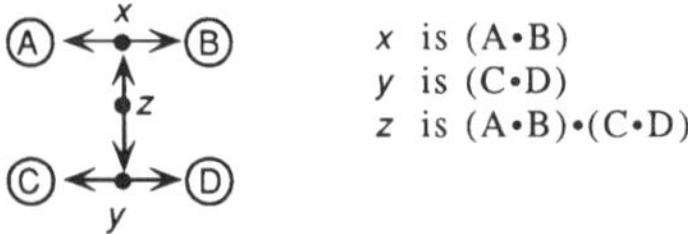

c. A state consisting of two separate dimers.

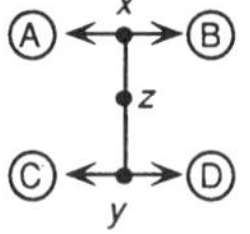

z is a state in which A is bound to B and C is bound to D, but the 2 dimers may not be bound to each other.

d. A and B bind independently to different domains on C, forming a trimer that can be considered to be the union of two dimers.

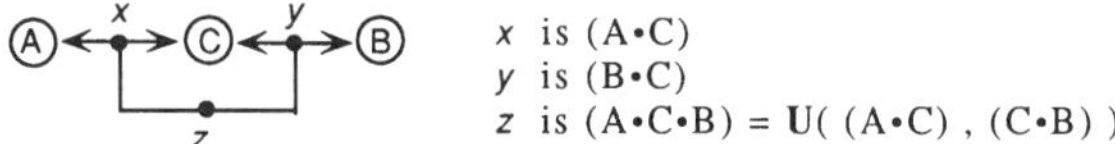

FIGURE 7.3 Symbol definitions. (A) Symbols used in the explicit diagram in Figure 7.2. (B) Additional symbols used in the heuristic diagram in Figure 7.4.

cyclin E and E2F has been added to the system. The feedback synthesis reaction (reaction 23) preserves the timed E2F wave and increases the magnitude of the E2F response.

The final component to be added is cyclin D–Cdk4 which acts on Rb and E2F similarly to cyclin E–Cdk2, except that cyclin D is generated in a wave by external control and there is no feedback loop. In order to see the effect the cyclin D stimulus by itself, the effect of the cyclin E–dependent kinase can be blocked by setting $k_{15} = 0$ (corresponding to $b = 0$ in Figure 7.1A). The result is an early and limited wave of E2F. When both cyclin D and cyclin E activities are included in the simulation, the E2F response is larger and more prolonged, and is followed by an increase in cyclin E–Cdk2. This behavior is consistent with our previous simulations

B

Additional symbols used in heuristic diagrams

Stimulation via transcription (direct or indirect)

Catalytic stimulation of a reaction (*eg* by enzyme or transcription factor)

Inhibition of transcription

Inhibitions of all other kinds

Bond cleavage (as by a phosphatase)

Examples:

e. A and B bind mutually exclusively to C, as by competing for the same domain on C.

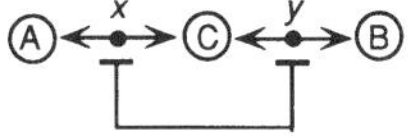

f. Complex A•B catalyzes the phosphorylation of a site on C

FIGURE 7.3 (continued)

using the differential equations (Figure 7.1). Thus, cyclin D can serve as an early trigger of a larger and more prolonged activation of cyclin E leading to S phase. In the absence of cyclin D, however, cyclin E can nevertheless eventually become activated at a later time.

It is interesting to note that generally similar behavior can be obtained by constitutive expression of cyclin E, without the feedback control of the promoter by E2F. The role of the feedback control may be to provide a large E2F response over a wider range of conditions. Thus, an important aspect in the evolution of a control network would be to provide increasing robustness of control.

The behavior of cyclin D, cyclin E, Rb, and E2F in these simulations is consistent with what has been observed experimentally, and has properties that are consistent with what would be expected of a network that functions in control of the G1/S phase transition. Moreover, the network can be viewed as having evolved stepwise from simpler networks capable of more primitive functions. The experimental simulator can, in this sense, attempt to recapitulate evolution. Indeed, an understanding of how a complex control network functions may be achieved by asking how the network might have evolved.

7.4 HEURISTIC DIAGRAMS

In order to proceed with simulations of increasing complexity, it is important to have an overview of what is known about the molecular interactions that may be implicated in the network. Information on molecular interactions operating in biological regulatory networks is rapidly accumulating. The need to integrate this information in the form of functional models is increasingly urgent, but the task is made difficult by the complexity of the interactions being revealed.[17] Functional models are frequently represented by diagrams, which can help comprehension of complex systems, but no consistent convention has developed by which diagrams could be interpreted unambiguously. Diagrams of the kinds commonly presented in molecular biology papers and talks often suffer form two deficiencies: (1) inability to cope with highly complex systems and (2) ambiguity in excess of the data — it is often difficult to deduce exactly what is meant without extensive reading of accompanying text. The complexity of the emerging interaction networks is such that prediction and interpretation of experimental results are becoming increasingly difficult and may require computer simulations.

Ideally, it would be useful to have a well-defined convention of molecular interaction diagrams that would be concise and not limited by complexity. A molecular interaction diagram may be considered to be "well defined" if it can be uniquely translated into a file readable by a simulation program where only the rate constants and initial conditions remain unspecified. However, there is a trade-off between well-defined computer-oriented diagrams and those that might readily be comprehended by humans. Moreover, computer programs require all of the interactions to be precisely specified, while many details are often unknown. Well-defined diagrams often require many *ad hoc* assumptions which would be made for the purpose of specific trial simulations. Therefore, it is also useful to develop diagrams that, albeit not entirely well defined as far as computers are concerned, nonetheless present the known essentials in an integrated and comprehensible manner: heuristic diagrams would allow ambiguity commensurate with the uncertainty of knowledge. Heuristic diagrams can be employed to select explicit well-defined subsystems for simulation trials.

Regulatory networks differ from classical metabolic pathway maps in several ways:

1. The enzymes, such as kinases and phosphatases, are often substrates of other enzymes.
2. A substrate may in large part be bound to one or more of its effector enzymes.
3. Functional species often consist of multiprotein complexes, the formation and dissociation of which can be integral to network function.
4. Interactions at the gene level are highly complex and may introduce time delays.

A diagram convention for regulatory networks must cope with these features.

Definition of Symbols

The symbols used in the well-defined explicit diagram (Figure 7.2) are defined in Figure 7.3A, and additional symbols that will be needed for the heuristic diagram in Figure 7.4A are explained in Figure 7.3B.

The first consideration for symbol selection was ease of representation of protein–protein binding and multiprotein complexes. Binding between two molecular species is represented by a double-arrowed line with barbed arrowheads. One double-arrowed line can point to another, thus allowing complexes to be built up indefinitely. To define a particular complex, a small filled circle is placed on the line. These definitions are illustrated in Figure 7.3A, examples a and b. It is sometimes necessary to define a state consisting of more than one independent molecular species; a method for doing this is illustrated in Figure 7.3A, example c. A convenient way to refer to different subsets of a multimer is suggested in Figure 7.3A, example d; this method is especially useful in explicit diagrams. The use of a mutual exclusivity symbol is illustrated in Figure 7.3B, example e. Catalysis of a reaction is represented by a small open circle (Figure 7.3B, example f). Strong or covalent binding of a small molecule or group (e.g., phosphorylation) is indicated by a single-arrowed line with barbed arrowhead (Figure 7.3B, example f). Additional rules are (1) multiple filled circles within the same line (excluding the ends of the line) refer to the same species and (2) binding lines (terminated by barbed arrows at each end) can change direction, but cannot branch.

7.5 A HEURISTIC DIAGRAM OF THE CONTROL NETWORK GOVERNING THE G1/S TRANSITION

Figure 7.4 organizes the known or suspected molecular interactions in the control network involving G1 and S phase cyclin-dependent kinases and E2F transcription factors. The numbers in the chart refer to references listed in Figure 7.4B. Question marks are applied to processes that are in doubt or for which there is conflicting evidence.

Some of the organizing elements in the diagram will be mentioned. Three parallel pathways, controlled, respectively, by cyclins D, E, and A, are displayed horizontally. Whereas cyclins E and A both operate through Cdl2 and p107, cyclin D operates exclusively through Cdk4/6 and Rb. Cyclin E–Cdk2 also may operate via Rb, and probably has other important substrates that remain to be elucidated. The E2F transcription factors are heterodimers (e.g., E2F1–DP1) that are controlled by members of the Rb/p107/p130 family which in turn are controlled by the cyclin-dependent kinases. Cyclin A–Cdk2 could have the special function of reducing or terminating E2F activity at the end of S phase by phosphorylating one or both of its monomer components, such as E2F1 and DP1. The cyclin-dependent kinases are controlled by members of the p21/p27 family; p21 additionally binds Gadd45 and PCNA and thereby may directly affect DNA replication or repair. p53 functions in part by activating the transcription of p21, Gadd45, and PCNA. Another level of control, specifically on cyclin D–Cdk4/6 is mediated by p16 and p15 which compete with cyclin D for binding to Cdk4/6. The diagram also shows the manner in which S phase genes may be controlled jointly by the E2Fs and Sp1. Included among the E2F-regulated genes are the genes for several components of the control network itself, thus providing multiple opportunities for feedback circuits (for example, involving cyclin E and c-myc) with time delays that could make the function of the network

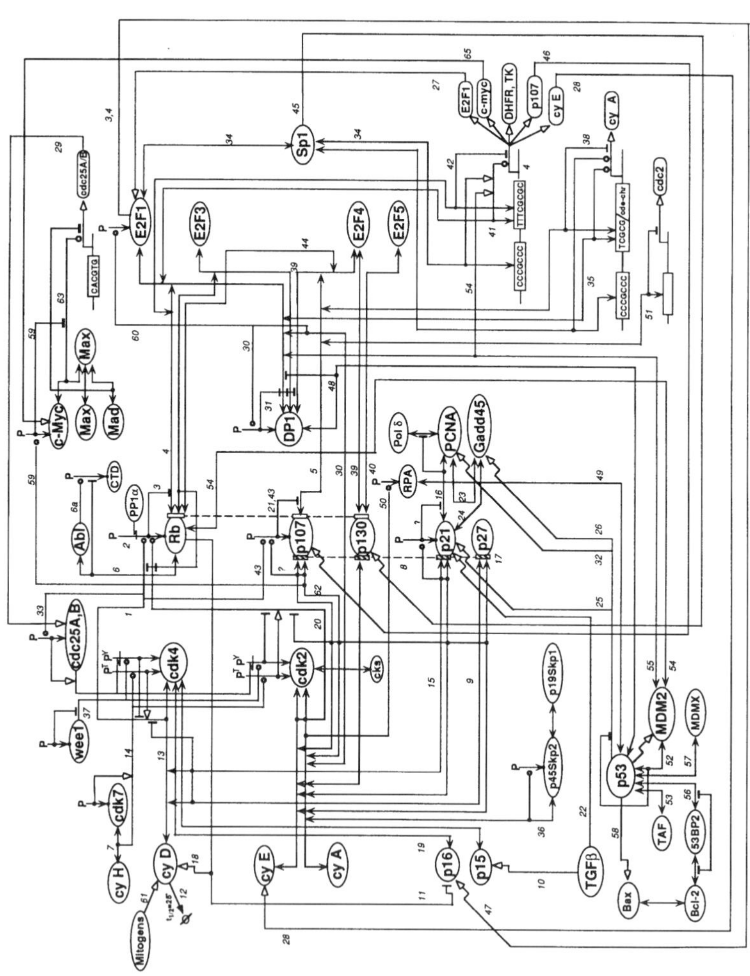

difficult to predict and that may favor instabilities. These connections can be traced on the diagram (Figure 7.4A) and literature references can be looked up in Figure 7.4B.

As new information accumulates, the diagram will continually have to be updated. This could be done interactively via the internet. The localization of molecular species within and transfer between intracellular compartments may be an important aspect for addition to the diagram as this information becomes available.

7.6 PERSPECTIVES

A wealth of information about the basic molecular components responsible for cell cycle control is at present available and continues to accumulate at a rapid pace. This is a fruitful base for further development of more-sophisticated mathematical models applicable to mammalian cells, similar to those already developed for yeasts by Tyson and colleagues. Mathematical modeling of the cell cycle may not only help in better understanding of the logic in controlling the cell cycle, but also may suggest new experiments for identification of new components and relationships important for cell cycle regulation.

It is a daunting task to organize and keep track of the accumulating information on the molecular interactions needed for this modeling. The task may be aided by suitable diagram conventions such as the method we have applied to G1/S phase regulation. Such diagrams might achieve clarity comparable with electric circuit diagrams. We have shown that two types of diagrams are needed: explicit diagrams that precisely define a system for simulation and heuristic diagrams that show the general nature of the molecular interactions. It should be noted that our diagrams do not contain specific reference to biological events; such events may be viewed as emergent behavior that should arise out of the molecular interactions.

FIGURE 7.4 (A) Heuristic diagram of molecular interactions known or suspected to be involved in the control of the G1/S phase transition and the progress of S phase. For symbol definitions, see Figure 7.3. Proteins are indicated within ovals, promoter elements within rectangles, and mRNAs within capsule shapes. Binding domains on proteins may be indicated by hatched rectangles fused to an oval protein symbols, and homology of domains on different proteins may be indicated by connecting them with a dashed line (e.g., the homologous binding domains indicated for p21, p27, p107, and p130. Adjoining parallel binding arrows pointing to the same protein or domain are assumed to bind competitively to the same site. Binding to either of two related species is sometimes denoted by two barbed arrows at the same end of a line (as in the binding of p21 to cyclin E–Cdk2 or cyclin A–Cdk2). An action by either of two or more related species is sometimes conveniently indicated by superimposing the lines emerging from each species (as in the proposed phosphorylation of p21 by cyclin D–Cdk4, cyclin E–Cdk2, or cyclin A–Cdk2). CTD, C-terminal domain of RNA polymerase II; PP, protein phosphatase; cy, cyclin. (B) References for G1/S control network heuristic diagram (italic numbers refer to reaction numbers in A).

1 Kato JY, Matsuoka M, Polyak K, et al. (1994) Cell 79: 487-96
Ewen ME, Sluss HK, Sherr CJ, et al. (1993) Cell 73: 487-97
Matsushime H, Ewen ME, Strom DK, et al. (1992) Cell 71: 323-34
2 Hinds PW, Mittnacht S, Dulic V, et al. (1992) Cell 70: 993-1006
3 Chellappan SP, Hiebert S, Mudryj M, et al. (1991) Cell 65: 1053-61
4 Nevins JR (1992) Science 258: 424-9
Weinberg RA (1995) Cell 81: 323-30
LaThangue NB (1994) TIBS 19: 108-114
Ginsberg D, Vairo G, Chittenden T, et al. (1994) Genes Dev 8: 2665-79
Beijersbergen RL, Kerkhoven RM, Zhu L, et al. (1994) Genes Dev 8: 2680-90
Lees JA, Saito M, Vidal M, et al. (1993) Mol Cell Biol 13: 7813-25
6 Welch PJ and Wang JYJ (1993) Cell 75: 779-790
Welch PJ and Wang JYJ (1995) Genes Dev 9: 31-46
Welch PJ and Wang JYJ (1995) Mol Cell Biol 15: 5542-51
6a Duyster J, Baskaran R and Wang JYJ (1995) Proc Natl Acad Sci U S A 92: 1555-9
7 Fisher RP and Morgan DO (1994) Cell 78: 713-24
Makela TP, Tassan JP, Nigg EA, et al. (1994) Nature 371: 254-7
8 Toyoshima H and Hunter T (1994) Cell 78: 67-74
9 Slingerland JM, Hengst L, Pan CH, et al. (1994) Mol Cell Biol 14: 3683-94
Polyak K, Lee MH, Erdjument-Bromage H, et al. (1994) Cell 78: 59-66
Kato JY, Matsuoka M, Polyak K, Massague J and Sherr CJ (1994) Cell 79: 487-96
10 Hannon GJ and Beach D (1994) Nature 371: 257-61
11 Li Y, Nichols MA, Shay JW, et al. (1994) Cancer Res 54: 6078-82
Bates S, Parry D, Bonetta L, et al. (1994) Oncogene 9: 1633-40
Sherr CJ and Roberts JM (1995) Genes Dev 9: 1149-63
12 Parry D, Bates S, Mann DJ, et al. (1995) Embo J 14: 503-11
Bates S, Parry D, Bonetta L, et al. (1994) Oncogene 9: 1633-40
13 Meyerson M and Harlow E (1994) Mol Cell Biol 14: 2077-86
Bates S, Bonetta L, MacAllan D, et al. (1994) Oncogene 9: 71-9
14 Kato JY, Matsuoka M, Strom DK, et al. (1994) Mol Cell Biol 14: 2713-21
Matsuoka M, Kato JY, Fisher RP, et al. (1994) Mol Cell Biol 14: 7265-75
15 Harper JW, Elledge SJ, Keyomarsi K, et al. (1995) Mol Biol Cell 6: 387-400
Chen IT, Akamatsu M, Smith ML, et al. (1996) Oncogene 12: 595-607
16 Zhang H, Hannon GJ and Beach D (1994) Genes Dev 8: 1750-8

Luo Y, Hurwitz J and Massague J (1995) Nature 375: 159-61
Flores-Rozas H, Kelman Z, Dean FB, et al. (1994) Proc Natl Acad Sci U S A 91: 8655-9
Chen IT, Akamatsu M, Smith ML, et al. (1996) Oncogene 12: 595-607
17 Nakanishi M, Robetorye RS, Adami GR, et al. (1995) Embo J 14: 555-63
18 Bates S, Parry D, Bonetta L, et al. (1994) Oncogene 9: 1633-40
Muller H, Lukas J, Schneider A, et al. (1994) Proc Natl Acad Sci U S A 91: 2945-9
19 Parry D, Bates S, Mann DJ and Peters G (1995) Embo J 14: 503-11
Koh J, Enders GH, Dynlacht BD, et al. (1995) Nature 375: 506-10
Lukas J, Parry D, Aagaard L, et al. (1995) Nature 375: 503-6 (but see Hirai H, Roussel MF, Kato JY, et al. (1995) Mol Cell Biol 15: 2672-81)

FIGURE 7.4(B)

20 Faha B, Ewen ME, Tsai LH, et al. (1992) Science 255: 87-90
Zhu L, Enders G, Lees JA, et al. (1995) EMBO J 14: 1904-13
21 Beijersbergen RL, Carlee L, Kerkhoven RM, et al. (1995) Genes Dev 9: 1340-53
22 Reynisdottir I, Polyak K, Iavarone A, et al. (1995) Genes Dev 9: 1831-45
23 Smith ML, Chen IT, Zhan Q, et al. (1994) Science 266: 1376-80
Hall PA, Kearsey JM, Coates PJ, et al. (1995) Oncogene 10: 2427-33
Chen IT, Akamatsu M, Smith ML, et al. (1996) Oncogene 12: 595-607
24 Kearsey JM, Coates PJ, Prescott AR, et al. (1995) Oncogene 11: 1675-83
25 Kearsey JM, Coates PJ, Prescott AR, et al. (1995) Oncogene 11: 1675-83
26 Kastan MB, Zhan Q, el-Deiry WS, et al. (1992) Cell 71: 587-97
27 DeGregori J, Kowalik T and Nevins JR (1995) Mol Cell Biol 15: 4215-24
28 Duronio RJ and O'Farrell PH (1995) Genes Dev 9: 1456-68
Ohtani K, DeGregori J and Nevins JR (1995) Proc Natl Acad Sci U S A 92: 12146-50
Botz J, Zerfass-Thome K, Spitkovsky D, et al. (1996) Mol Cell Biol 16: 3401-9
30 Krek W, Ewen ME, Shirodkar S, et al. (1994) Cell 78: 161-72
Krek W, Xu G and Livingston DM (1995) Cell 83: 1149-58
Dynlacht BD, Flores O, Lees JA, et al. (1994) Genes Dev 8: 1772-86
Xu M, Sheppard KA, Peng CY, et al. (1994) Mol Cell Biol 14: 8420-31
31 Dynlacht BD, Flores O, Lees JA and Harlow E (1994) Genes Dev 8: 1772-86
Krek W, Ewen ME, Shirodkar S, et al. (1994) Cell 78: 161-72
32 Shivakumar CV, Brown DR, Deb S, et al. (1995) Mol Cell Biol 15: 6785-93
33 Ohtsubo M, Theodoras AM, Schumacher J, et al. (1995) Mol Cell Biol 15: 2612-24
34 Lin SY, Black AR, Kostic D, et al. (1996) Mol Cell Biol 16: 1668-75
Karlseder J, Rotheneder H and Wintersberger E (1996) Mol Cell Biol 16: 1659-67
35 Zwicker J, Lucibello FC, Wolfraim LA, et al. (1995) EMBO J 14: 4514-22
Schulze A, Zerfass K, Spitkovsky D, et al. (1995) Proc Natl Acad Sci U S A 92: 11264-8
36 Zhang H, Kobayashi R, Galaktionov K, et al. (1995) Cell 82: 915-25
37 Watanabe N, Broome M and Hunter T (1995) EMBO J 14: 1878-91
38 Smith EJ and Nevins JR (1995) Mol Cell Biol 15: 338-44
39 Sardet C, Vidal M, Cobrinik D, et al. (1995) Proc Natl Acad Sci U S A 92: 2403-7
Beijersbergen RL, Kerkhoven RM, Zhu L, et al. (1994) Genes Dev 8: 2680-90
Ginsberg D, Vairo G, Chittenden T, et al. (1994) Genes Dev 8: 2665-79
40 Hijmans EM, Voorhoeve PM, Beijersbergen RL, et al. (1995) Mol Cell Biol 15: 3082-9
41 Wu CL, Zukerberg LR, Ngwu C, et al. (1995) Mol Cell Biol 15: 2536-46
42 Helin K, Harlow E and Fattaey A (1993) Mol Cell Biol 13: 6501-8
43 Xiao ZX, Ginsberg D, Ewen M, et al. (1996) Proc Natl Acad Sci U S A 93: 4633-7
44 Ikeda MA, Jakoi L and Nevins JR (1996) Proc Natl Acad Sci U S A 93: 3215-20
45 Baldi A, Boccia V, Claudio PP, et al. (1996) Proc Natl Acad Sci U S A 93: 4629-32
46 Zhu L, Zhu L, Xie E, et al. (1995) Mol Cell Biol 15: 3552-62
47 Resnitzky D and Reed SI (1995) Mol Cell Biol 15: 3463-9
48 Sorensen TS, Girling R, Lee CW, et al. (1996) Mol Cell Biol 16: 5888-95
49 Dutta A, Ruppert JM, C. AJ, et al. (1993) Nature 365: 79-82
Li R and Botchan MR (1993) Cell 73: 1207-21
50 Dutta A and Stillman B (1992) EMBO J 11: 2189-99
51 Tommasi S and Pfeifer GP (1995) Mol Cell Biol 15: 6901-13
52 Kussie PH, Gorina S, Marechal V, et al. (1996) Science 274: 948-53

53 Thut CJ, Chen JL, Klemm R, et al. (1995) Science 267: 100-4
Lu H and Levine AJ (1995) Proc Natl Acad Sci U S A 92: 5154-8
Farmer G, Colgan J, Nakatani Y, et al. (1996) Mol Cell Biol 16: 4295-304
54 Martin K, Trouche D, Hagemeier C, et al. (1995) Nature 375: 691-4
55 Xiao ZX, Chen J, Levine AJ, et al. (1995) Nature 375: 694-8
56 Naumovski L and Cleary ML (1996) Mol Cell Biol 16: 3884-92
Gorina S and Pavletich NP (1996) Science 274: 1001-5
57 Shvarts A, Steegenga WT, Riteco N, et al. (1996) Embo J 15: 5349-57
58 Merchant AK, Loney TL and Maybaum J (1996) Oncogene 13: 2631-7
59 see Adams 1996 MCB 16:6623 (Intro).
60 Kitagawa M, Higashi H, Suzuki-Takahashi I, et al. (1995) Oncogene 10: 229-36
61 Lukas J, Bartkova J and Bartek J (1996) Mol Cell Biol 16: 6917-25
62 Xiao ZX, Ginsberg D, Ewen M and Livingston DM (1996) Proc Natl Acad Sci U S A 93: 4633-7
Zhu L, Harlow E and Dynlacht BD (1995) Genes Dev 9: 1740-52
63 Galaktionov K, Chen X and Beach D (1996) Nature 382: 511-7
64 Tiefenbrun N, Melamed D, Levy N, et al. (1996) Mol Cell Biol 16: 3934-44
65 Oswald F, Lovec H, Moroy T, et al. (1994) Oncogene 9: 2029-36

REFERENCES

1. Murray, A. and Hunt, T., *The Cell Cycle,* W.H. Freeman, New York, 1993.
2. Hutchison, C. and Glover, D. M., Eds., *Cell Cycle Control,* Oxford University Press, New York, 1995.
3. Nasmyth, K., Viewpoint: putting the cell cycle in order, *Science,* 274, 1643, 1996, and accompanying articles.
4. Reed, S. I. and Maller, J. L., Cell multiplication (Editorial overview), *Curr. Opin. Cell Biol.,* 8, 763, 1996, and accompanying articles.
5. Goldbeter, A., Modelling the mitotic oscillator driving the cell division cycle, *Biochemical Oscillations and Cellular Rhythms. The Molecular Bases of Periodic and Chaotic Behaviour,* Cambridge University Press, New York, 1996, chap. 10.
6. Goldbeter, A., Guilmot, J. M., and Romond, P. C., Control of cell proliferation: towards new strategies suggested by modelling the mitotic oscillator, *Pathol. Biol.* (Paris), 44, 172, 1996.
7. Goldbeter, A., A minimal cascade model for the mitotic oscillator involving cyclin and cdc2 kinase, *Proc. Natl. Acad. Sci. U.S.A.,* 88, 9107, 1991.
8. Novak, B. and Tyson, J. J., Numerical analysis of a comprehensive model of M-phase control in Xenopus oocyte extracts and intact embryos, *J. Cell Sci.,* 106, 1153, 1993.
9. Novak, B. and Tyson, J. J., Quantitative analysis of a molecular model of mitotic control in fission yeast, *J. Theor. Biol.,* 173, 283, 1995.
10. Duronio, R. J., Brook, A., Dyson, N., and O'Farrell, P. H., E2F-induced S phase requires cyclin E, *Genes Dev.,* 10, 2505, 1996.
11. Obeyesekere, M. N., Herbert, J. R., and Zimmerman, S. O., A model of the G1 phase of the cell cycle incorporating cyclin E/Cdk2 complex and retinoblastoma protein, *Oncogene,* 11, 1199, 1995.
12. Temin, H., Stimulation by serum of multiplication of stationary chicken cells, *J. Cell Physiol.,* 78, 161, 1971.

13. Pardee, A. B., A restriction point for control of normal animal proliferation, *Proc. Natl. Acad. Sci. U.S.A.*, 71, 1286, 1974.
14. Zetterberg, A., Larsson, O., and Wiman, K. G., What is the restriction point? *Curr. Opin. Cell Biol.*, 7, 835, 1995.
15. Weinberg, R. A., The retinoblastoma protein and cell cycle control, *Cell*, 81, 323, 1995.
16. Kholodenko, B. N. and Westerhoff, H. V., The macroworld versus the microworld of biochemical regulation and control, *TIBS*, 20, 52–54, 1995.
17. Loomis, W. F. and Sternberg, P. W., Genetic networks, *Science*, 269, 649, 1995.
18. Kohn, K. W., Functional capabilities of molecular network components controlling the mammalian G1/S cell cycle phase transition, submitted for publication, 1997.

8 Recent Advances in Modeling Stochastic Population Growth

Thomas R. Kiffe and James H. Matis

CONTENTS

Abstract — Deterministic and stochastic models of population growth play a key role in modern ecological theory. With improvements in computer technology, stochastic models are becoming as easy to work with on a practical level as deterministic models. A simple logistic model is discussed from both a deterministic and a stochastic point of view. Fundamental differences between the two models are illustrated in an example arising from our study of the Africanized honeybee. A new approach to stochastic population growth, one which deals directly with the mean, variance, etc., of the population distribution is presented.

8.1 INTRODUCTION

Population growth models constitute a fundamental part of modern ecological theory. Perhaps the most commonly used growth model is the classical Verhulst–Pearl logistic equation. Recent advances in computer technology, especially the availability of computer algebra programs, make it possible to develop and implement more-realistic growth models. In particular, it is now fairly straightforward to add stochastic effects to growth models and approximate the mean, variance, and skewness of a population. The stochastic model of population growth adds new insights to the deterministic model and also illustrates its limitations. The methods presented in this paper can be extended to complex growth models involving two or more interacting populations. For the sake of simplicity and clarity, we will only present

0-8493-7962-8/98/$0.00+$.50

a basic "density-dependent" stochastic model and illustrate the techniques by applying them to our study of the dispersal of the Africanized honeybee.

The implementation of the techniques presented here would not be feasible without combining computer algebra systems like Mathematica™ and traditional numerical analysis. The basic ideas are quite straightforward, but implementing them on a laboratory level requires the kind of tedious algebraic manipulation and calculation best done by a computer.

First, we shall quickly review the basic deterministic logistic growth model for a single population. Next, we shall formulate the stochastic version of this model and show how to obtain the exact distribution of the resulting stochastic process. Finally, we shall present an alternative approach to the stochastic model, one which computes the mean, variance, and skewness of the distribution directly. This new approach relies heavily on complex computer algebra calculations as well as standard numerical procedures for solving systems of nonlinear equations. An example will be discussed which illustrates some of the inadequacies of the deterministic model and the traditional techniques for dealing with the stochastic model.

8.2 DETERMINISTIC MODELS

A widely used model for describing population growth is the well-known logistic model. In this model one assumes that a population of size N has "per capita" birth and death rates given by

$$\lambda(N) = a_1 - b_1 N \tag{8.1}$$

and

$$\mu(N) = a_2 + b_2 N \tag{8.2}$$

respectively, where a_1, a_2, b_1, and b_2 are all nonnegative. The a_i are interpreted in population ecology as defining the "intrinsic" birth and death rates, and the b_i yield the "density-dependent" effects.[7] The population size, $N(t)$, satisfies the Verhulst–Pearl differential equation

$$N'(t) = (a_1 - a_2)N(t) - (b_1 + b_2)N^2(t) \tag{8.3}$$

whose solution is

$$N(t) = \frac{K}{1 + \left[(K - N_0)/N_0\right]\exp(-at)} \tag{8.4}$$

where $a = a_1 - a_2$, $b = b_1 + b_2$, $K = a/b$, and N_0 is the initial population size. K is interpreted as the "carrying capacity" of the population and is the asymptotic limit of $N(t)$.

8.3 STOCHASTIC POPULATION GROWTH

Although the deterministic model has a solution in closed form and is widely used to analyze population growth, it does have a major shortcoming. "Every real system must be considered to be subject to uncertainties of one type or another, all of which are ignored in the formulation of a deterministic model" (Gold,[3] p. 96). Stochastic models use birth–death processes to describe population growth. In this vein, consider the Bartlett et al.[2] model in which the population size, N, is a random variable. The probabilities of a single birth and of a single death for a population of size $N = n$, in an infinitesimal time interval Δt, are $\lambda_n \Delta t$ and $\mu_n \Delta t$, respectively, where

$$\lambda_n = \begin{cases} a_1 n - b_1 n^2, & \text{if } n < a_1/b_1 \\ 0, & \text{otherwise} \end{cases} \tag{8.5}$$

and

$$\mu_n = a_2 n + b_2 n^2 \tag{8.6}$$

The problem now facing us is to describe the distribution of the random population size $N(t)$ for each $t > 0$. There are two ways to obtain this distribution. One approach is to find the distribution directly; the other approach, which we have been developing, is to approximate the mean, variance, and skewness of the distribution of $N(t)$ for $t > 0$.

8.3.1 PROBABILITY-GENERATING FUNCTIONS

Let $p_n(t)$ denote the probability that $N(t) = n$. If L is the smallest integer greater than a_1/b_1, then $p_n(t) = 0$ for $n > L$ since, by Equation 8.5, the probability of a birth is zero for $n > L$ and $t > 0$. Hence, there are only a finite number of nonzero probability functions which describe the distribution of $N(t)$. These probability functions have been used by many researchers[6,8] to characterize the transient and "quasi-equilibrium" distributions of the process $N(t)$. The quasi-equilibrium distribution is a stochastic analogue of the "carrying capacity" obtained from the deterministic model.

The differential equations that the functions $p_n(t)$ must satisfy can be easily derived. The probability-generating function for the random process $N(t)$ is defined by

$$P(\theta, t) = \sum_{i=0}^{\infty} p_i(t)\theta^i \tag{8.7}$$

and it can be shown[1] that $P(\theta,t)$ satisfies the partial differential equation

$$\frac{\partial P}{\partial t} = (\theta - 1)(a_1\theta - a_2)\frac{\partial P}{\partial \theta} - \theta(\theta - 1)(b_1\theta + b_2)\frac{\partial^2 P}{\partial \theta^2} \tag{8.8}$$

Substituting Equation 8.7 into 8.8 and equating similar powers of θ, we get the system of ordinary differential equations

$$
\begin{aligned}
p_0'(t) &= \mu_1 p_1(t) \\
p_1'(t) &= \mu_2 p_2(t) - (\lambda_1 + \mu_1) p_1(t) \\
&\mathrm{M} \\
p_i'(t) &= \mu_{i+1} p_{i+1}(t) - (\lambda_i + \mu_i) p_i(t) + \lambda_{i-1}\, p_{i-1}(t) \\
&\mathrm{M} \\
p_L'(t) &= -\mu_L p_L(t) + \lambda_{L-1}\, p_{L-1}(t)
\end{aligned}
\tag{8.9}
$$

Under the assumption that the initial population size is j, the initial conditions are $p_j(0) = 1$ and $p_i(0) = 0$ for $i \neq j$.

This system of equations, although linear, does not have a closed form solution. However, one can obtain the probabilities by solving the system numerically. Once the probabilities $p_n(t)$ have been calculated, one can compute the mean value function $m(t)$ and the variance function $\sigma^2(t)$ by

$$m(t) = \sum_{i=1}^{L} i p_i(t) \tag{8.10}$$

and

$$\sigma^2(t) = \sum_{i=1}^{L} i^2 p_i(t) - m^2(t) \tag{8.11}$$

Computer algebra programs and standard numerical procedures allow us to solve for the probabilities and hence completely describe the distribution of the random population size. There is a serious limitation to this approach, however. If L is large, it may be impossible to solve the equations numerically due to numerical instability. Hence, this approach is limited to populations of relatively small size. To overcome this limitation, we have developed a new approach to the stochastic population model by working with the mean, variance, etc., of the population directly.

8.3.2 Cumulant Functions

For any distribution $N(t)$ the uncorrected moments, $m_i(t)$, are defined by

$$m_i(t) = E\left[N^i(t)\right] = \sum_{n=0.}^{\infty} n^i p_n(t) \tag{8.12}$$

and the moment generating function $M(\theta, t)$ of the distribution of $N(t)$ is defined by

$$M(\theta,t) = E\left[e^{\theta N(t)}\right] = \sum_{i=1}^{\infty} m_i(t)\frac{\theta^i}{i!} \tag{8.13}$$

The cumulants, $\kappa_i(t)$, of the distribution are a widely used transformation of the moments, and they are defined through the relationship

$$m_i = \sum_{j=0}^{i-1} \binom{i-1}{j} \kappa_{i-j} m_j$$

with $m_0 = 1$.[4] In particular, the first three cumulants are

$$\begin{aligned} \kappa_1 &= m_1 \\ \kappa_2 &= m_2 - m_1^2 \\ \kappa_3 &= m_3 - 3m_1m_2 + 2m_1^3 \end{aligned} \tag{8.14}$$

and these are just the mean, variance, and skewness of the distribution of $N(t)$. Our new methodology involves solving for the cumulant functions rather than the probability functions. This approach is independent of the maximum population size and hence is applicable to very large populations.

The cumulant generating function $K(\theta, t)$ of the process $N(t)$ is just the logarithm of the moment-generating function. Under Assumptions 8.5 and 8.6, the cumulant-generating function must solve the nonlinear partial differential equation[1]

$$\frac{\partial K}{\partial t} = \left[\left(e^{\theta}-1\right)a_1 + \left(e^{-\theta}-1\right)a_2\right]\frac{\partial K}{\partial \theta} + \left[-\left(e^{\theta}-1\right)b_1 + \left(e^{-\theta}-1\right)b_2\right]\left[\frac{\partial^2 K}{\partial \theta^2} + \left(\frac{\partial K}{\partial \theta}\right)^2\right] \tag{8.15}$$

The cumulant generating can be written in terms of the cumulants κ_i by

$$K(\theta,t) = \sum_{i=1}^{\infty} \kappa_i(t)\frac{\theta^i}{i!} \tag{8.16}$$

If one substitutes this series expansion into Equation 8.15 and expands in powers of θ, one obtains the differential equations for the cumulant functions. Without the aid of computer algebra programs this computation can be extremely tedious. The equations for any number of cumulants may be derived in this fashion. For the first three cumulants we get

$$\begin{aligned} \kappa_1'(t) &= \left(a - b\kappa_1\right)\kappa_1 - b\kappa_2 \\ \kappa_2'(t) &= \left(c - d\kappa_1\right)\kappa_1 + \left(2a - d - 4b\kappa_1\right)\kappa_2 - 2b\kappa_3 \\ \kappa_3'(t) &= \left(a - b\kappa_1\right)\kappa_1 + \left(3c - b - 6d\kappa_1 - 6b\kappa_2\right)\kappa_2 + \left(3a - 3d - 6b\kappa_1\right)\kappa_3 - 3b\kappa_4 \end{aligned} \tag{8.17}$$

where

$$a = a_1 - a_2, \quad b = b_1 + b_2, \quad c = a_1 + a_2, \quad d = b_1 - b_2$$

8.4 AN EXAMPLE

We will illustrate these techniques with an example arising from our research on the dispersal of the Africanized honeybee.[5] In the deterministic formulation of the dispersal we assume that the population size, $N(t)$, is governed by the following assumptions:

1. Population birth rate = $0.30N - 0.015N^2$.
2. Population death = $0.02N + 0.001N^2$.

The differential equation for $N(t)$ becomes

$$N'(t) = 0.28N(t) - 0.016N^2(t) \tag{8.18}$$

and the solution, with $N(0) = 2$, is the logistic growth function

$$N(t) = \frac{17.5}{1 + 7.75 \ \exp(-0.28t)} \tag{8.19}$$

The carrying capacity is 17.5 and the maximum population size is 20.

In the stochastic formulation we suppose that the random population size $N(t)$ has the following unit change probabilities in the interval Δt:

3. Probability of a single birth = $(0.30N - 0.015N^2)\ \Delta t$.
4. Probability of a single death = $(0.02N + 0.001N^2)\ \Delta t$.

The differential equations for the first three cumulants are

$$\kappa_1'(t) = 0.28\kappa_1 - 0.016\kappa_1^2 - 0.016\kappa_2 \tag{8.20}$$

$$\kappa_2'(t) = 0.32\kappa_1 - 0.014\kappa_1^2 + 0.546\kappa_2 - 0.064\kappa_1\kappa_2 - 0.032\kappa_3 \tag{8.21}$$

$$\begin{aligned}\kappa_3'(t) = {} & 0.28\kappa_1 - 0.016\kappa_1^2 + 0.944\kappa_2 - 0.084\kappa_1\kappa_2 - 0.096\kappa_2^2 + 0.798\kappa_3 \\ & - 0.096\kappa_1\kappa_3 - 0.048\kappa_4\end{aligned} \tag{8.22}$$

Since $\kappa_1(t)$ and $\kappa_2(t)$ are just the mean and variance of the random process $N(t)$, comparing Equations 8.18 and 8.20 we see that the stochastic model reduces to the deterministic model if we assume that the variance is zero. The deterministic solution must always overestimate the mean of the population since κ_2, the variance, is inherently positive. Figures 8.1, 8.2, and 8.3 show the graphs of the mean, variance

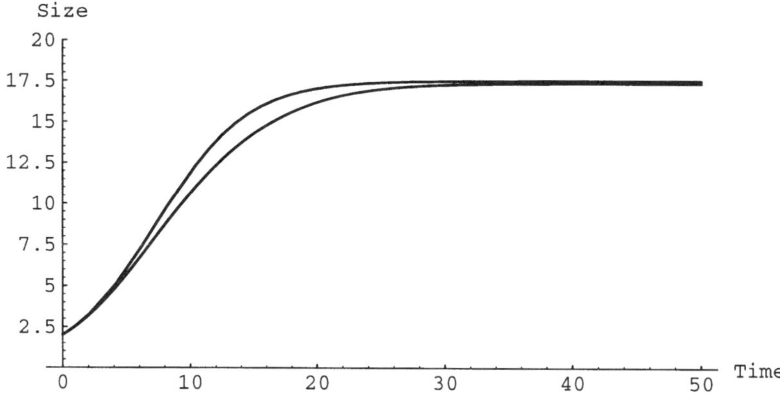

FIGURE 8.1 The plot of the deterministic solution and the mean function for the Africanized honeybee. Note that the deterministic solution lies above the mean function.

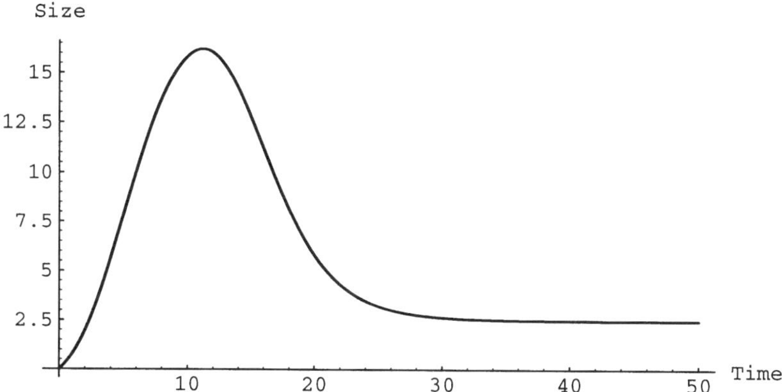

FIGURE 8.2 The plot of the variance function for the Africanized honeybee. Note the relatively large variance during the initial period of growth.

and skewness of the population. These graphs are obtained by setting κ_4 to zero and solving Equations 8.20, 8.21, and 8.22 numerically.

8.5 DETERMINISTIC VS. STOCHASTIC MODELS

We conclude this chapter with an example from Renshaw[8] which illustrates the dramatic differences that can arise between deterministic and stochastic models. Under Assumptions 8.1 and 8.2 the deterministic model always produces a population function which has asymptotic value as time increases. This asymptotic value is called the *carrying capacity* of the environment and is easily calculated to be $(a_1 - a_2)/(b_1 + b_2)$. The question naturally arises as to whether the stochastic model under Assumptions 8.5 and 8.6 has an asymptotic distribution. Suppose that such a distribution, X, does exist and let q_n be the probability that $X = n$. Since

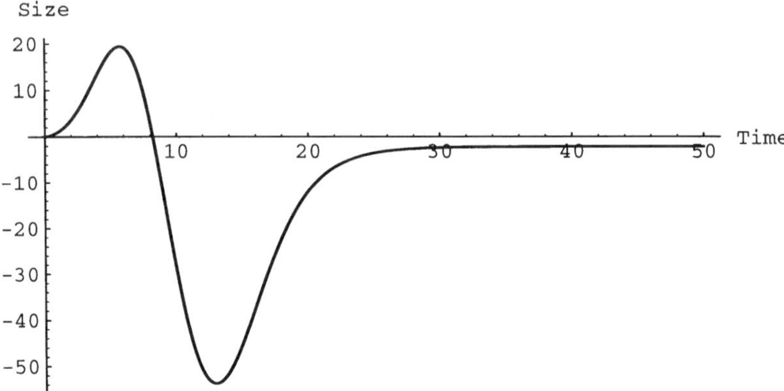

FIGURE 8.3 The plot of the skewness function for the Africanized honeybee. The skewness of the process changes dramatically during the initial period of growth.

$$\lim_{t \to \infty} p_n(t) = q_n \tag{8.23}$$

taking the limit in Equation 8.9 gives the system

$$\begin{aligned}
0 &= \mu_1 q_1 \\
0 &= \mu_2 q_2 - (\lambda_1 + \mu_1) q_1 \\
&\text{M} \\
0 &= \mu_{i+1} q_{i+1} - (\lambda_i + \mu_i) q_i + \lambda_{i-1} q_{i-1} \\
&\text{M} \\
0 &= -\mu_L q_L + \lambda_{L-1} q_{L-1}
\end{aligned} \tag{8.24}$$

The only solution to this system is $0 = q_1 = q_2 = \ldots$ so that $q_0 = 1$; i.e., the asymptotic population size is zero with probability one. In other words, ultimate extinction of the population is certain. Ecologists have been aware of this fact for decades. However, Bartlett et al.[2] observe that a quasi-equilibrium distribution "may effectively exist for all realizable time-intervals" for most processes of ecological interest. Renshaw[8] (pp. 61–69) discusses an example of a stochastic growth process under the Assumptions 8.5 and 8.6 with $a_1 = 2.2$, $a_2 = 0.2$, $b_1 = 0.1$, and $b_2 = 0.1$. The deterministic model with these parameters is

$$N'(t) = 2.0N(t) - 0.2N^2(t) \tag{8.25}$$

and the solution, with $N(0) = 2$, is the logistic growth function

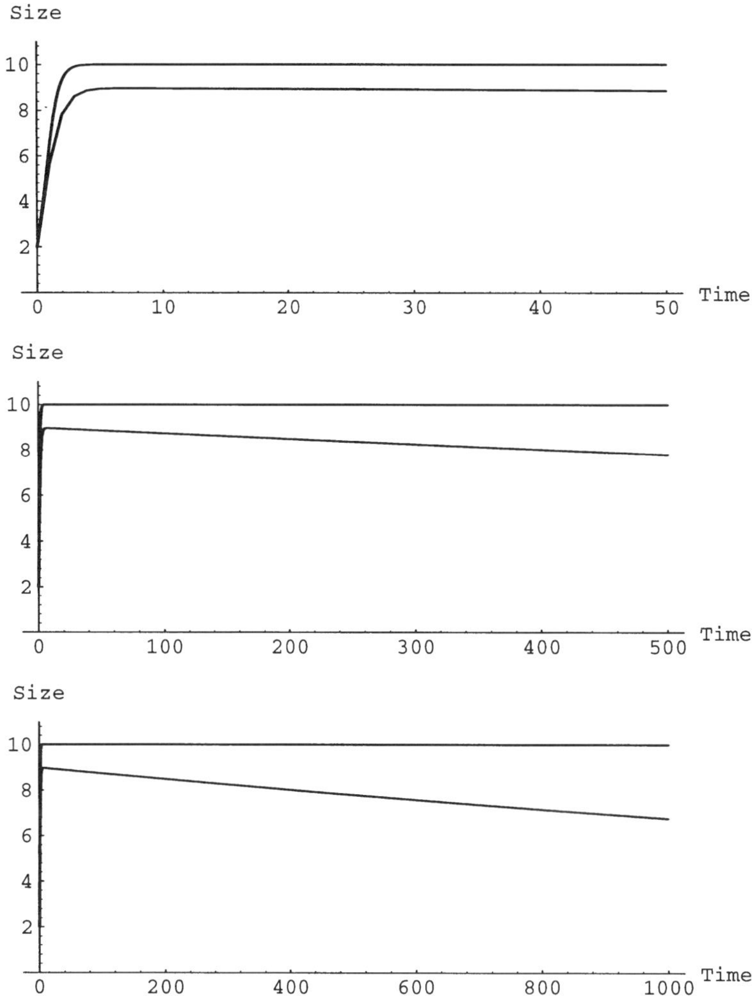

FIGURE 8.4 The plot of the deterministic solution and the mean function for the Renshaw example. Note that the deterministic solution lies above the mean function and note the increasing difference between the deterministic solution and the mean function.

$$N(t) = \frac{10}{1 + 4\ \exp(-2t)} \tag{8.26}$$

The carrying capacity is 10 and the maximum population size is 22. The deterministic model predicts that the population size will approach 10 as an asymptotic limit. The stochastic model gives very different results. Substituting these parameter values for a_1, a_2, b_1, and b_2 into Equation 8.9 with $L = 22$ and solving numerically, we can find exactly the true distribution for this process. From the discussion above we know that the mean function $m(t)$ approaches zero as t approaches infinity and Figure 8.4

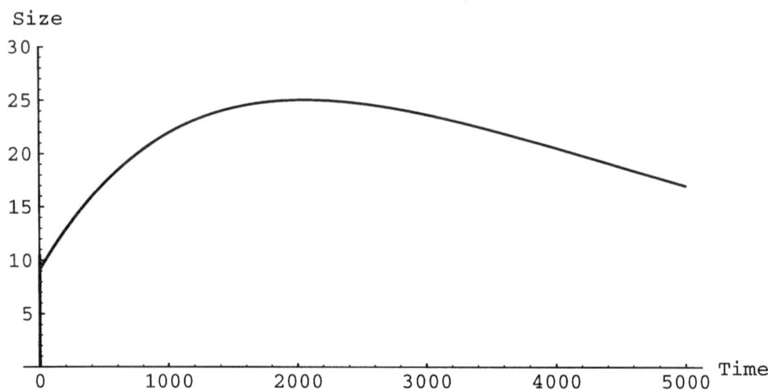

FIGURE 8.5 The plot of the variance function for the Renshaw example. Recall that the quasi-equilibrium distribution gives an equilibrium variance of 6.6.

illustrates how substantially the deterministic model overestimates the true population mean, even for reasonable values of time.

It is not our intention to discuss the quasi-equilibrium distribution here other than to illustrate, using Renshaw's example, that the assumption of such a distribution may be misleading. Renshaw shows that the mean time to extinction of the process is about 2972 years and estimates the probability of extinction by $t = 10{,}000$ to be over 0.96. These results are consistent with Figure 8.4 and raise the question of the appropriateness of the quasi-equilibrium assumption. As an illustration, Renshaw shows the mean of the quasi-equilibrium solution to be 9.29, while our results give a maximum value of 8.97 at time $t = 7$ and the mean has already decreased to 8.7 by time $t = 100$. Figure 8.5 illustrates that the behavior of the true variance is radically different than the predicted value of 6.6 from the quasi-equilibrium distribution. The point is that one can always calculate a quasi-equilibrium distribution, conditional upon the process not being extinct; however, such distributions may be very different, even for moderate "realizable" time intervals, from the exact (unconditional) distribution illustrated in this paper.

REFERENCES

1. Bailey, N. T. J., *The Elements of Stochastic Processes,* John Wiley, New York, 1964.
2. Bartlett, M. S., Gower, J. C., and Leslie, P. H., A comparison of theoretical and empirical results for some stochastic population models, *Biometrika,* 47, 1–11, 1960.
3. Gold, H., *Mathematical Modeling of Biological Systems,* John Wiley, New York, 1977.
4. Johnson, N. L., Kotz, S., and Kemp, A. W., *Univariate Discrete Distributions,* 2nd ed., John Wiley, New York, 1992.
5. Matis, J. H. and Kiffe, T. R., On approximating the moments of the equilibrium distribution of a stochastic logistic model, *Biometrics,* 52, 980–991, 1966.
6. Nisbet, R. M. and Gurney, W. S. C., *Modeling Fluctuating Populations,* John Wiley, New York, 1982.
7. Odum, E. P., *Basic Ecology,* Saunders, New York, 1983.
8. Renshaw, E., *Modeling Biological Populations in Space and Time,* Cambridge University Press, New York, 1991.

9 A Reconstructability Analysis Approach to Biological Systems Modeling

S. K. Trivedi, B. Jones, and S. S. Iyengar

CONTENTS

Abstract — Living systems essentially represent complex information-processing systems that are governed by numerous biological and natural processes. These processes are often nonstationary, and exhibit chaotic behavior. It is difficult to model such processes, or systems governed by such processes. The black box approaches enable us to study input/output patterns, disregarding cumbersome internal details of living

0-8493-7962-8/98/$0.00+$.50

systems. However, it is important to study the interactions and interrelationships among system variables in order to gain a precise understanding of such systems.

The concepts of system reconstruction and system identification in systems theory offer promising tools for analyzing complex biological and natural systems. Reconstructability theory relates to two types of problems. The first is referred to as the reconstruction problem which deals with the process of reconstructing a given system under a given criterion from the knowledge of its subsystems and, during this process, identifying those subsystems that are vital to the reconstruction. The second is referred to as the identification problem which allows the identification of an unknown system from the knowledge of its subsystems. The solution procedures associated with these two problems are referred to as reconstructability analysis. In this chapter, we introduce the basic terminology and concepts in reconstructability analysis and discuss how reconstructability analysis is adequately suitable for modeling of complex biological and natural systems. We also present case studies relating to the problems in ecology and plant pathology.

9.1 INTRODUCTION

Life is essentially an information-processing system governed by a number of biological and natural processes. For example, mutation can be explained as a sudden change in the information-processing content of the hereditary substance. Also, viruses can be regarded as carriers of complex hereditary substances with very high information contents. Biological processes are often nonstationary, complex, and manifold, and usually involve a large number of variables.[2,9,43] Behavior of these processes is mostly chaotic. The underlying equations for such processes are extremely complicated and impossible to solve. For this reason most biological and natural systems are treated as black boxes. This enables us to study input/output patterns without paying much attention to the cumbersome internal mechanisms of such systems. However, in order to gain precise understanding of these systems, it is important to study the interactions and interrelationships among system variables. The concepts of system reconstruction and system identification in systems theory offer promising tools for analyzing complex biological and natural systems.

Reconstructability theory relates to two types of problems, namely, the reconstruction problem and the identification problem. The former deals with the process of reconstructing a given system under a given criterion from the knowledge of its subsystems and, during this process, identifying those subsystems that are vital to the reconstruction. The latter allows the identification of an unknown system from the knowledge of its subsystems. The solution procedures associated with these two problems are referred to as reconstructability analysis, abbreviated as RA. Origins of RA can be traced to Ashby's[1] work on constraint analysis in the early 1960s,[29] although a formal framework of RA did not exist until the late 1970s[3,25,26] and the early 1980s.[4-7] Since the advent of reconstructability analysis, researchers have directed significant efforts in this area. This has resulted in the emergence of a variety of new algorithms and applications. Solution procedures aimed at these two problems have been developed and implemented.[13,14,16-31,35,36,44-46] In this chapter, based on Reference 5, we introduce the basic terminology and concepts in RA. We shall discuss how RA is adequately suitable for modeling of complex biological and

TABLE 9.1
An Example of a Probabilistic Behavior System

v_1	v_2	v_3	f
0	0	0	0.079
0	0	1	0.088
0	0	2	0.083
0	1	0	0.031
0	1	1	0.052
0	1	2	0.097
1	0	0	0.091
1	0	1	0.072
1	0	2	0.037
1	1	0	0.109
1	1	1	0.128
1	1	2	0.133

natural systems and present case studies relating to the problems in ecology and plant pathology.[10,37]

9.2 PRELIMINARY CONCEPTS IN RECONSTRUCTABILITY THEORY

9.2.1 Systems and States

Intuitively, a system is simply a data set which consists of the tuples of the form $\langle v_1, v_2, \ldots, v_n, f \rangle$, where $v_1, v_2, \ldots, v_n$ are variables or attributes and f is a function defined over these variables. This function may be a probability distribution function, a possibilistic behavior function, a selection function, a fuzzy set membership function, or any arbitrary or nonlinear function. A state in a system is simply a combination of attribute values in a given order. Formally, a system is defined as a six tuple

$$B = (V, W, s, A, Q, f) \tag{9.1}$$

where $V = \{v_i \mid i \in 1, 2, \ldots, n\}$ is a set of variables; $W = \{V_j \mid j \in \{1, 2, \ldots, m\}\ m \le n\}$ is a family of state sets; s: $V \rightarrow W$ is an onto mapping which assigns to each variable in V one state set from W; $A = s(v_1) \times s(v_2) \times \cdots \times s(v_n)$ is the set of all potential aggregate states; Q is a set of real numbers; and f: $A \rightarrow Q$ is a system function which represents the information regarding the aggregate states of the system. Note that we refer to B and f as system and system function, respectively, instead of behavior system and behavior function as was originally the case.[5] This generalization is evident from the fact that RA methodology has evolved to cover functions other than behavior functions.[21-24] An example of a probabilistic behavior system is shown in Table 9.1. In this example, $V = \{v_1, v_2, v_3\}$ is the set of variables; $W = \{\{0,1\}, \{0,1,2\}\}$ is a family of state sets; s: $V \rightarrow W$ is an onto mapping which assigns {0,1} to v_1 and v_2, and {0,1,2} to v_3; A = {000, 001, 002, 010, 011, 012, 100, 101, 102, 110, 111, 112}; Q = [0,1] is a set of real numbers and f: $A \rightarrow Q$ is a probabilistic behavior function.

9.2.2 Subsystems and Substates

A subsystem is a data set whose variables form a proper subset of the variables of the system and a function g is defined over variables in the subset. It consists of the tuples of the form $\langle u_1, u_2, \ldots, u_m, g\rangle$, where $\{u_1, u_2, \ldots, u_m\} \subset \{v_1, v_2, \ldots, v_n\}$. A substate in a subsystem is a combination of attribute values present in that subsystem in a given order. We can regard a system as a subsystem of a larger system and a subsystem as a system (i.e., a supersystem) of a smaller system. Formally, given a system as defined in Equation 9.1, a collection of a total of q subsystems, together called a structure system or a reconstruction hypothesis, is defined as

$$S = \{{}^k B\} = \{({}^k V, {}^k W, {}^k s, {}^k A, {}^k Q, {}^k f) \,|\, k \in \{1, 2, \ldots, q\}\} \tag{9.2}$$

if and only if, for each k the following conditions are satisfied:

1. ${}^kV \subset V$,
2. ${}^kW \subseteq W$ such that ${}^ks$ is onto,
3. ${}^ks$: ${}^kV \rightarrow {}^kW$ such that ${}^ks(v_i) = s(v_i)$ for each $v_i \in {}^kV$,
4. ${}^kA = \times_{v_i \in {}^kV} {}^ks(v_i)$,
5. ${}^kQ = Q$,
6. ${}^kf = [f \downarrow {}^kV]$.

Elements of set S are referred to as subsystems of system B. $[f \downarrow {}^kV]$ is called the projection of f on ${}^kV$, which considers only the variables in ${}^kV$. Essentially, $[f \downarrow {}^kV]$ is a mapping from substates in ${}^kA$ to Q, that is,

$$[f \downarrow {}^k V]: \underset{v_i \in {}^k V}{\times} s(v_i) \rightarrow Q \tag{9.3a}$$

such that

$$[f \downarrow {}^k V](\beta) = g(\{f(\alpha) \,|\, \alpha > \beta\}) \tag{9.3b}$$

where $\alpha > \beta$ means β is a substate of α (or α is a superstate of β), and g is determined by the nature of function f. For instance, if f is a probabilistic distribution function, then

$${}^k f(\beta) = \sum_{\alpha > \beta} f(\alpha) \tag{9.4a}$$

TABLE 9.2
An Example of Typical Subsystems of a Probabilistic Behavior System

v_1	v_2	${}^{12}f$	v_2	v_3	${}^{23}f$	v_1	v_3	${}^{13}f$	v_1	${}^{1}f$	v_2	${}^{2}f$	v_3	${}^{3}f$
0	0	0.25	0	0	0.17	0	0	0.11	0	0.43	0	0.45	0	0.31
0	1	0.18	0	1	0.16	0	1	0.14	1	0.57	1	0.55	1	0.34
1	0	0.20	0	2	0.12	0	2	0.18					2	0.35
1	1	0.37	1	0	0.14	1	0	0.20						
			1	1	0.18	1	1	0.20						
			1	2	0.23	1	2	0.17						

or if f is a selection or possibilistic behavior function, then

$$ {}^{k}f(\beta) = \max_{\alpha>\beta} f(\alpha). \tag{9.4b}$$

An example of typical subsystems of the previously defined system is shown in Table 9.2. In this example, a reconstruction hypothesis or structure system can be described as $S = \{{}^{k}B\} = \{\{v_1,v_2\}, \{v_1,v_3\}, \{v_2,v_3\}, \{v_1\}, \{v_2\}, \{v_3\}\}$. Illustratively, ${}^{12}V = \{v_1,v_2\}$ is the set of variables; ${}^{12}W = \{\{0,1\}\}$ is a family of state sets; ${}^{12}s$: ${}^{12}V \rightarrow {}^{12}W$ is an onto mapping which assigns $\{0,1\}$ to v_1 and v_2; ${}^{12}A = \{{}^{12}(00), {}^{12}(01), {}^{12}(10), {}^{12}(11)\}$; ${}^{12}Q = [0,1]$ is a set of real numbers; and ${}^{12}f$: ${}^{12}A \rightarrow {}^{12}Q$ is a probabilistic behavior function. Similarly, ${}^{3}V = \{v_3\}$ is the set of variables; ${}^{3}W = \{\{0,1,2\}\}$ is a family of state sets; ${}^{3}s$: ${}^{3}V \rightarrow {}^{3}W$ is an onto mapping which assigns $\{0,1,2\}$ to v_3; ${}^{3}A = \{{}^{3}(0), {}^{3}(1), {}^{3}(2)\}$; ${}^{3}Q = [0,1]$ is a set of real numbers; and ${}^{3}f$: ${}^{3}A \rightarrow {}^{3}Q$ is a probabilistic behavior function. Note that we have used projection functions to derive subsystems. This may not be valid in the real world where different subsystems are observed by different observers by the means of different experiments. This will require resolution of local and global inconsistencies.[5]

9.2.3 Reconstruction Problem

Let B be a behavior system defined by Equation 9.1. Let S be a structure system defined by Equation 9.2. S is said to be a meaningful reconstruction hypothesis of B if and only if it contains the subsystems of B, such that

$$ \bigcup_{k\in N_q} {}^{k}V = V \tag{9.5}$$

and

$$ \left(\text{for all } j,k \in N_q\right)\left({}^{j}V \subseteq {}^{k}V \Rightarrow j = k\right). \tag{9.6}$$

Condition 9.5 is called the *covering condition* which guarantees that all variables of B are included in S. This means that the reconstruction of B from S is logically possible. Condition 9.6 is called the *irredundancy condition* which ensures that S contains no redundant information.

9.2.4 Identification Problem

Let S be a structure system defined by Equation 9.2. Then S is said to be a reconstruction hypothesis or hypothetical representation of an unknown overall system B provided the following six conditions hold true:

1. $V = \bigcup_{k \in N_q} {}^kV$,
2. $W = \bigcup_{k \in N_q} {}^kW$,
3. s: $V \rightarrow W$ such that $s(v_i) = {}^ks(v_i)$ for each $k \in N_k$,
4. $A = \times_{v_i \in V} s(v_i)$,
5. $Q = {}^kQ$ for each $k \in N_q$,
6. $f{:}A \rightarrow Q$ such that $[f \downarrow {}^kV] = {}^kf$ for each $k \in N_q$.

It is required that B be compatible with S. In fact, there are more than one overall systems compatible with S. The set of all these systems is called *reconstruction family* of S, denoted by B_s. As discussed previously, if the behavior functions $\{{}^kf\}$ of a reconstruction hypothesis are projections of an overall behavior function f, then the reconstruction hypothesis is *consistent*. A reconstruction hypothesis is locally consistent if the following condition, called the *local consistency condition*, is satisfied:

$$\left(\text{for all } j,k \in N_q\right)\left(\left[{}^jV \downarrow {}^jV \cap {}^kV\right] = \left[{}^kV \downarrow {}^jV \cap {}^kV\right]\right). \tag{9.7}$$

A reconstruction hypothesis S is *globally consistent* if the reconstruction family of S is nonempty. In most cases, reconstruction hypotheses satisfy the local consistency condition 9.7 as well as the covering condition 9.5 and the irredundancy condition 9.6.

9.2.5 Evaluation of Reconstruction Hypotheses

Let B be a given system.[5] Then there exists a family of meaningful reconstruction hypotheses ζ_B, for B. Given a particular reconstruction hypothesis S in this family, there exists B_s, a family of systems which are compatible with the hypothesis. A system is compatible if the projections of the system function are included in that particular reconstruction hypothesis as subsystem functions. B_s is called the reconstruction family of S. The term reconstruction uncertainty is used to express the size of the reconstruction family. This defines an identifiability quotient in order to identify a unique system from a given reconstruction hypothesis. Making a choice of a single system function f_s from the reconstruction family B_s requires some assumptions in order to justify the particular choice. The most significant theoretical justification[5] is that the determined system function f_s should be non-committal in all regards except in satisfying

$$\left[f_s \downarrow {}^k V\right] = {}^k f = \left[f \downarrow {}^k V\right] \quad (9.8)$$

for all $k \in \{1,2, \ldots, q\}$. For a probabilistic system function f, the above condition implies that the set of values $\{f_s(\alpha) \mid \alpha \in A\}$ must have the maximum entropy among all such sets associated with the systems in B_s. This implies that the reconstruction must be a maximum entropy solution subject to Condition 9.8. In the context of a reconstruction problem, as stated in Reference 5, the principle of maximum entropy can be justified by the following arguments:

1. The maximum entropy reconstruction is the only unbiased reconstruction as it takes into account all available information but no additional information.[5]
2. The maximum entropy reconstruction is the most likely reconstruction.[5]
3. Maximizing any function other than entropy will lead to inconsistencies except when that function has the same maxima as entropy.[42]
4. A system is a real-world system if and only if it has the maximum entropy reconstruction. In other words, it is impossible to design a real-world system whose actual reconstruction is different from the maximum entropy reconstruction.[5]

For probabilistic systems, Shannon's[41] entropy is used as the measure of uncertainty which satisfies the properties of symmetry, expansibility, subadditivity, additivity, normalization, and continuity.[27,41] For a possibilistic system, Shannon's entropy is replaced by U-*uncertainty*[13,27] which satisfies the property of monotonicity in addition to the properties satisfied by its probabilistic counterpart. Therefore, the above arguments also hold good for justifying the principle of maximum uncertainty reconstruction in the context of possibilistic systems.[7,14,30]

The problem of determining the unbiased reconstruction f_s for a reconstruction hypothesis S is essentially an optimization problem.[5] Given the set of functions h such that $h: A \rightarrow Q$, determine f_s for which the entropy $-\Sigma_{\alpha \in A}\, h(\alpha) \log h(\alpha)$ reaches its maximum subject to Condition 9.8. Cavallo and Klir[5] provided a solution to this problem. Later, Cavallo and Klir,[7] Higashi et al.,[14] and Klir and Parviz[30] extended RA to cover the possibilistic structure systems and possibilistic system functions using the principle of maximum uncertainty.

9.3 K-SYSTEMS

Jones[16-18] received his motivation from Lewis's study[32] on approximation of probability distributions to reduce storage requirements. He revolutionized the field of RA in that he extended the system/subsystem paradigm of RA to the state/substate paradigm by modifying RA to work at the levels of states and substates and to work for general nonlinear functions.[16-24] He provided alternative methods of solutions using minimal, incomplete, and arbitrary information in the context of reconstruction and identification problems by introducing the concepts of null extension and K-systems theory. The concept of null extension and the advent of K-systems have significantly extended the reach of RA methodology. By transforming any nonlinear system to a dimensionless system, it is possible for RA to work for most nonlinear functions. By using the concept of null extension, it is possible to divide the whole state space into disjoint equivalent classes and to proceed further by just picking only one state from each class.

Jones formalized the concepts of a G-system and a K-system. Let R^+ be the set of positive real numbers. Let f and τ be defined as follows:

$$f: A \rightarrow R^+, \tag{9.9a}$$

$$\tau = \sum_{\alpha \in A} f(\alpha). \tag{9.9b}$$

Then, a G-system is defined as the following tuple:

$$\left(\tau, \{v_i\}, \{\alpha\}, \{\beta\}, f, \{^m f\}\right). \tag{9.10}$$

Here τ is a parameter of interest to define a K-system, $\{v_i\}$ is a set of variables, $\{\alpha\}$ is a set of states, $\{\beta\}$ is a set of substates, f is a function on $\{\alpha\}$, and $\{^m f\}$ are functions on $\{\beta\}$. Now, in order to define a K-system, we do the following transformation. We define

$$k(\alpha) = f(\alpha)/\tau, \quad \text{for all } \alpha \tag{9.11a}$$

so that

$$0 \le k(\alpha) \le 1, \quad \text{for all } \alpha, \tag{9.11b}$$

and

$$\sum_{\alpha} k(\alpha) = 1. \tag{9.11c}$$

Now, a K-system is defined as the following tuple:

$$\left(\tau, \{v_i\}, \{\alpha\}, \{\beta\}, k, \{^m k\}\right) \tag{9.12}$$

where τ is a transformation factor, $\{v_i\}$ is a set of variables, $\{\alpha\}$ is a set of states, $\{\beta\}$ is a set of substates, k is a function on $\{\alpha\}$, and $\{^m k\}$ are functions on $\{\beta\}$. From this description, it is obvious that a K-system is a dimensionless system,

TABLE 9.3
Human Temperature Phenomenon as a G-System

$X_1\ X_2\ X_3\ X_4 \ldots X_n$	$f(X_1, X_2, X_3, X_4, \ldots, X_n)$
Observations of variables	Observations of temperature

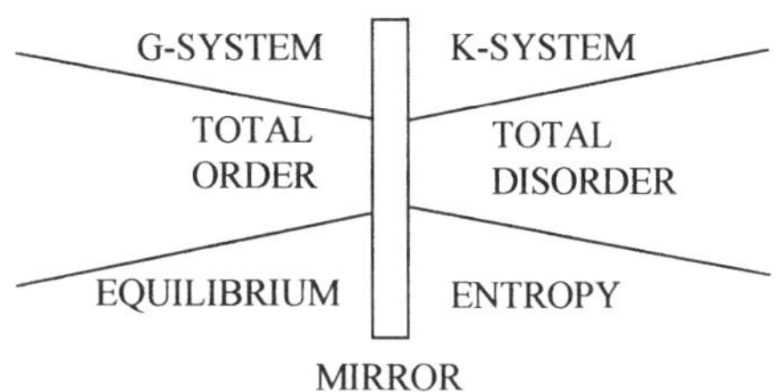

FIGURE 9.1 G-system vs. K-system.

although it is isomorphic to a G-system. There is no loss or gain of information in switching from one system to the other. The goal of K-systems methodology is to utilize the framework of probabilistic RA in order to develop solution procedures for G-systems.

9.3.1 G-Systems vs. K-Systems

Let us consider the normal human temperature which is 98.6°F. We may view this as a state of order and, thus, as a state of equilibrium. We can study this phenomenon as a G-system as shown in Table 9.3. Now, consider mapping this G-system into a K-system. The state of equilibrium of the G-system corresponds to the state of maximum entropy of the K-system. Thus, we have two different functional representations, namely, the G-system function and the K-system function. In fact, we are measuring order in two ways, equilibrium and entropy. The state of maximal order in the G-system corresponds to the state of maximal disorder in the K-system, giving a mirror effect. This is shown in Figure 9.1 and is true for any G-system.

It is clear that the G-function is left invariant by K-system transformation. In particular, this is true for any probabilistic G-system. We note that for the same system function, the state of maximal order or equilibrium is the state of maximal disorder or entropy. This is in agreement with the fact that all known equilibrium results of quantum mechanics are based on the principle of maximum entropy.

9.3.2 Significance of RA and K-Systems

RA deals primarily with systems and subsystems,[5] more specifically, with states and substates[16-24] and the interrelationships among them. The state/substate paradigm of RA has led to a departure from the classical statistical approach, has initiated the design of more-powerful algorithms and, thus, has provided new insights into the structure and dynamics of systems. Contrary to RA, classical statistical approaches emphasize variables.[23] In RA, an event is the occurrence of a state or substate, but an event is not a state or substate by itself. A state exists regardless of its occurrence, and states can be referred to without implying their occurrence. Thus, RA has the concept of independent events, which has a classical equivalent, and the concept of independent states, which has no classical equivalent. The concept of independent states has played a significant role in the design of algorithms and procedures in RA. A state viewed apart from its occurrence is important in RA, illustratively, in null extensions[16-24] or in evaluating the cognitive contents of system substrates.[20] Earlier maximum entropy algorithms designed by Jones were employed successfully only after the advent of RA.

The inherent advantage of the RA methodology, more specifically, of K-systems, over the classical techniques is that it does not assume any structure in the data. RA is as good as the data it considers. On the contrary, classical techniques induce some kind of model on the data and introduce extraneous information. For instance, the analysis of variance, one of the most powerful techniques in the field of statistical inference, assumes a linear model on the data.[23] In short, in standard statistical analysis the true behavior of the system is masked by the overall effects, while K-systems analysis allows variables to act and interact in combinations.

9.4 K-SYSTEMS ANALYSIS

This section introduces important concepts in K-systems analysis which, in fact, is an extension of RA. K-systems analysis is concerned with the phenomena or systems from which observations are taken. It uses special sums to measure effects of system states and substates on the system behavior. This effectively eliminates the possibility of unnecessary constraints on the data. For instance, suppose we study a plant under different conditions and record the data as given in Table 9.4.

Humidity, leaf area, and disease are variables, and the corresponding numbers in the table are variable values. Yield loss is a system function and its corresponding numbers in the table are called system function values. Each line in the table constitutes an observation. We are interested in investigating this data to find the important factors or states or substates that affect yield loss. An example of a factor is (Humidity = 40, Disease = 20). Another factor would be (Humidity = 60). A factor is formed by taking specific values for any subset or combination of variables.

Jones[24] developed a program for performing K-systems analysis on data sets by concentrating information into a few descriptive factors or substates. The analysis proceeds by creating categories or clusters for any variable which is not already categorized. The analysis employs a hierarchical clustering technique which uses a minimum distance criterion to form centroids and then looks for a natural breakpoint in the centroids. K-systems analysis is applicable for categorical as well as continuous data.[24] When the values of a variable do not fall into discrete categories, we must cluster the data. Each cluster value represents a range of variable values. Clustering coarsens the independent variable values into categories, but does not affect the dependent variable. In fact, clustering simply consolidates a number of variable values into representation by a single value, and allows the user to view

TABLE 9.4
Study of Yield Loss in a Plant as a Function of Humidity, Leaf Area, and Disease

Humidity	Leaf Area	Disease	Yield Loss
60	4	0	0.0
60	4	10	1.0
60	4	20	8.0
60	5	0	0.0
60	5	20	8.0
60	6	0	0.0
60	6	20	8.0
50	4	0	2.0
50	4	20	12.2
50	5	0	1.5
50	5	20	11.7
50	6	0	1.0
50	6	10	2.8
50	6	20	11.2
40	4	0	5.0
40	4	20	16.5
40	5	0	3.0
40	5	10	5.1
40	5	20	14.5
40	6	0	1.0
40	6	10	3.1
40	6	20	12.5

effects at different levels of comprehensiveness. Empirical and theoretical evidence indicates that three clusters per variable are often effective in resolving system behavior. Intuitively, if a crop gets too much or too little rain, it perishes; but in between the crop flourishes. Composition of each factor is determined by entropy mathematics.

Moreover, the system function may be incomplete; that is, $f(\alpha)$ is unknown for certain α. In order to counter this problem the analysis assigns the mean of the known function values to each unknown function value. This process is referred to as entropy fill and results in minimizing the information added to the subsequent K-systems where the $f(\cdot)$ function has been transformed to a probabilistic equivalent (0,1) function. Also, in practice it is possible that there are two or more function values for the same state as a result of clustering of variables. In this case, we average the redundant values to obtain a single value for a state. Now, we can invoke information theoretic algorithms.

The transformation to a K-system gives a dimensionless (0,1) system which can be analyzed using entropy mathematics. Such an analysis yields structural information which is isomorphic under this simple mapping to the original G-system. This concludes the front end of the K-systems analysis. In summary, we

1. Cluster values of continuous variables;
2. Average redundant system function values;
3. Entropy-fill missing system function values; and
4. Map the original system into a K-system.

Now, we have values for all $k(\alpha)$ and, therefore, we can compute all ${}^{m}k(\beta)$. We proceed by forgetting all $k(\alpha)$. Using only the knowledge of substates ${}^{m}k(\beta)$, we determine which of the substates contributes the most to our knowledge of $k(\alpha)$. We add substates one by one to the set of important substates until this set adequately explains the behavior of $k(\alpha)$ for the whole system. Let $k'(\alpha)$ denote a current approximation to $k(\alpha)$, and let ${}^{m}k'(\beta)$ be an approximate value derived from appropriate $k'(\alpha)$. The analysis finds substates or factors that most influence the system behavior one by one in order of importance using the following procedure.

1. Initialize $k'(\cdot)$ to a flat distribution in terms of a (0,1) function with all values receiving the same weight.
2. Let $\Omega(\cdot)$ be a choice function defined subsequent to this algorithm. Now select the next β such that $\beta = \text{argmax } \Omega(\beta)$ over β not yet selected.
3. Compute a new $k'(\cdot)$ as $k'(\cdot) = \mu(\{\beta \mid \text{all } \beta \text{ selected}\})$ where $\mu(\cdot)$ is defined subsequent to this algorithm.
4. Ask if $k'(\cdot)$ is sufficiently close to $k(\cdot)$. That is, ask if

$$100 * \left[1 - \left(\Sigma k(\alpha) \log_2 \left(k(\alpha)/k'(\alpha)\right) / \Sigma k(\alpha) \log_2 \left(k(\alpha)/k''(\alpha)\right)\right)\right] \geq \delta \qquad (9.13)$$

If yes, then stop, else go to step (2).

Here $k'(\alpha)$ is an approximation to the true $k(\alpha)$. $k''(\alpha)$ is a flat distribution, and δ is a prescribed tolerance. Note that the flat distribution is an initial distribution which contains no information. Formula 9.13 is referred to as the system accuracy and represents a measure of information distance which is more accurate than the relative error measure from numerical analysis. The systems accuracy represents the degree to which a set of factors captures the total information contained in the system function. The strength of a factor is measured by the extent of information it contributes to the reconstruction of the original system.

Now, we define $\Omega(\cdot)$ and $\mu(\cdot)$. Let

$$I\left[{}^{m}k(\beta), {}^{m}k'(\beta)\right] = {}^{m}k(\beta) \log_2 \left({}^{m}k(\beta) / {}^{m}k'(\beta)\right)$$

then

$$\Omega(\beta) = I\left[{}^{m}k(\beta), {}^{m}k'(\beta)\right] + I\left[\left(1 - {}^{m}k(\beta)\right), \left(1 - {}^{m}k'(\beta)\right)\right]. \qquad (9.14)$$

By using this formula, the analysis picks the β that will add the most information to the system under reconstruction. The unbiased reconstruction or maximum entropy approximation of $k(\cdot)$ from the selected states or substates, $\mu(\cdot)$, is computed as follows. With each selected β, we can associate an equation

$$^{m}k(\beta) = \Sigma k(\alpha) \quad \text{over} \quad \alpha \geq \beta \tag{9.15}$$

so that the selected set of β forms a set of linear equations in which $k(\alpha)$ are unknown. We can rewrite this set of linear equations as

$$\sum_{j} k_{ij} = a_i \quad \text{for all} \quad a_i = ^{m} k(\beta). \tag{9.16}$$

We solve these equations using the following iterative algorithm.

(1) Initialize $k(\cdot)$ to a flat distribution.
(2) For all i: new $k_{ij} = k_{ij}(a_i/a'_i)$ for every j where a'_i is derived from the current estimate of k_{ij} and a_i is the true value.
(3) Convergence test: Is | new k_{ij} – old k_{ij} | $\leq \delta$ for all ij? If satisfied then stop, else go to step 2.

This completes the determination of the set of important system factors or substates. Additionally, the analysis offers the ability to compute interactions and to predict the effects of factors not present in the original data. This requires the notion of a factor effect. An effect is a change in the system behavior or system function value, expressed relative to a flat system behavior or the mean system behavior. An interaction is defined as the contribution to a factor effect that is due to the components of a factor acting all in unison.

Let $V = \{v_1 = c_1, v_2 = c_2, \ldots, v_n = c_n\}$ be a set of n variables with a particular value assignment to each variable, that is, a factor or a substate. Let V_m be the set of all subsets of V, and let V_P denote the set of all proper subsets of V. We compute the unbiased reconstructions $\mu(V_m)$ and $\mu(V_P)$. Now, for any factor which embodies V, we consider the difference in its effect as computed from these two unbiased reconstructions. This difference represents the contribution to a factor effect that is due to the components acting in unison. In order to predict the effect of a factor that is not present in the original data, a process very similar to the computation of an interaction is performed. Even when the factor effect is unknown, we can often form accurate estimates of some of its effects of subfactors. The unbiased reconstruction can now be computed from such subfactors. The resulting distribution provides a prediction of the effect for the overall factor. Note that the goodness of such a prediction will depend on the importance of any missing subfactor interactions and the missing overall interaction. However, such a prediction does make use of all the known relevant information.

During K-systems analysis, it is important to know how widespread is the effect associated with a factor. This is referred to as the factor region size. On clustering a variable, many values become represented by a single value. This single value, in fact, represents an interval or a simple region of variable values. A factor is a combination of such single values. Thus, a factor actually represents many factors which we refer to as a region. The number of factors which a factor actually represents relative to the total number of values available to be represented by all factors is called the region size of that factor.

We invoked the K-systems analysis on the data given in the preceding example. This revealed the following results. The analysis augmented the data with a dummy variable. This allowed the factors of all n variables to occur in the analysis. The extent to which possible combinations of the variable values are present in the data is referred to as system completeness. This is a measure of the information quantity present in the data. Controlling factors refer to the dominant factor necessary to explain or reproduce the total system behavior to the accuracy specified. The isolated effect measures how a factor by itself increases or decreases the behavior of a system. The intermediate effect displays factors that modify the effect of a chosen factor when they occur conjointly with the factor. The complete factor search refers to the search over all possible factors. Clustering was performed prior to the analysis. K-systems analysis gave the output which is shown in Table 9.5. The list of controlling factors, in the order of their importance, is shown in Table 9.6.

The analysis revealed that the variable disease occurred in five factors, humidity occurred in four factors, and leaf area occurred in two factors. The factor (Disease = 20.00) had an intermediary effect of –29.82% by the factor (Humidity = 60.00). The factor (Disease = 20.00) also had an intermediary effect of –7.31% by the factor (Leaf area = 6.00). For the factor (Humidity = 60.00, Leaf area = 4), –23.66% of the total effect was due to the factor interaction effect.

9.5 K-SYSTEMS ANALYSIS APPLIED TO ECOLOGICAL DATA: A CASE STUDY

In ecological systems the relationships among environmental factors are well defined at times, such as relationship between photosynthesis and light, or are arbitrary, such as relationship between predator and prey. The standard classical techniques are not capable of modeling such systems without introducing extraneous or erroneous information during the process. Shaffer et al.[37-40] have conducted various ecological case studies using K-systems analysis. This section is based on a study conducted by Shaffer and Cahoon[37] in the eastern arm of Mugu Lagoon, Ventura County, CA. K-systems analysis was performed under muddy and nonmuddy conditions to study benthic microfloral productivity; to determine the minimal set of independent variables in order to account for most of the dynamic variation in system function benthic microfloral productivity; and to decide how this set affected the overall behavior of the system. Variables taken into consideration were light, benthic community respiration, subaerial exposure, and mean tidal range. Additional variables were the ratio of pheophytin-a to chlorophyll-a, which represents the physiological state of microflora, water temperature next to the sediment cores, and initial dissolved oxygen.

TABLE 9.5
K-Systems Analysis Output of the Yield Loss Data

Cluster Map for Humidity
Values: 40.00 to 45.00 are represented by cluster value: 40.00.
Values: 45.00 to 55.00 are represented by cluster value: 50.00.
Values: 55.00 to 60.00 are represented by cluster value: 60.00.

Cluster Map for Leaf Area
Values: 4.00 to 4.50 are represented by cluster value: 4.00.
Values: 4.50 to 5.50 are represented by cluster value: 5.00.
Values: 5.50 to 6.00 are represented by cluster value: 6.00.

Cluster Map for Disease
Values: 0.00 to 5.00 are represented by cluster value: 0.00.
Values: 5.00 to 15.00 are represented by cluster value: 10.00.
Values: 15.00 to 20.00 are represented by cluster value: 20.00.

State resolution is 100%.
Maximum resolution offset is 0.00.

The data is 81.48% complete as a system description.
Analysis can proceed for any level of completeness, and results will be correct.

Special note for clustered data

To obtain a higher level of completeness, you must recluster one or more of the variables into a smaller number of cluster values. To resolve the states more precisely, you must increase the number of clusters or add a variable — results are correct for the cluster level that is employed; the system is just being discerned more coarsely when resolution is low.

Analysis created a complete set of factors.
Analysis searched the entire factor set.
Analysis will find the factors that describe 90.00% of the behavior of yield loss up to a maximum of 20 factors.

Results
Number of controlling factors = 6.
Number of distinct variables in controlling factors = 3.
System accuracy of controlling factors = 92.6%.
Average error for the reproduced system = 0.43.
Maximum error for the reproduced system = 3.94.

K-systems analysis indicated that the physiological state of microflora and initial dissolved oxygen had highly inconsistent effects on productivity. These variables were therefore excluded from the analysis.

About 1000 samples were incubated during a single month. A bidaily set of samples included within-site variability of benthic microfloral production and standing crop. Comparisons of four analytical procedures, namely, stepwise regression, time series, multichannel information, and K-systems analysis, were made. K-systems analysis was found to be more accurate as compared with other techniques.

TABLE 9.6
List of Important Factors for the Yield Loss Data

Factor	Isolated Effect on Flat System, % (Effect Value)	Region Size	System Accuracy after Adding This Factor (%)
1. Disease = 20.000	95.8 (11.40)	25.00	45.95
2. Disease = 0.000	–74.2 (1.50)	25.00	71.93
3. Humidity = 60.000	–30.1 (4.07)	25.00	80.18
4. Humidity = 60.000 Disease = 0.000	–100.0 (0.00)	6.25	86.83
5. Humidity = 60.000 Leaf area = 4.000 Disease = 10.000	–82.8 (1.00)	3.13	89.65
6. Humidity = 40.000 Leaf area = 4.000 Disease = 0.000	–14.1 (5.00)	1.56	92.57

9.5.1 Methodology and Analysis

Considering the difficulties involved in performing an accurate analysis of benthic microfloral production because of dissimilarities in productivity and standing crop, the results achieved using K-systems analysis were found to be more comprehensive. As a result of this study, it was discovered that K-systems analysis is appropriate for most ecological data because of the sporadic changes in the data caused by nonlinear events.[37] Conducting standard statistical analysis required elimination or transformation of part of the data in order to suit the underlying model. On the contrary, K-systems modeled nonlinearities in the data without such transformation. This is because K-systems analysis imposes no structure on the data that is not explicitly present in the data and uses factors or states (or substates) which are more general than variables. A true picture of ecological system dynamics was achieved without regard to complicated interrelationships among system variables. Table 9.7 displays important factors discovered during the analysis.[37]

In particular, the analysis concluded that the productivity increased with low tidal range, high standing crop, and high light, and decreased, in general, with high tidal range, low standing crop, and low light. Analysis showed that subaerial exposure increased the productivity for a short period. This was followed by a decrease in productivity due to overexposure. Subaerial exposure accounted for the lack of agreement between the biweekly cycles in productivity at the two tidal stations. Additionally, there existed a cyclic pattern of microfloral production. This was mainly induced by the disturbances caused by the spring tides. Benthic microflora were able to respond quickly to the integrated dynamic effects of other variables in the system. This was due to their high turnover rates and capability of storing ample nutrient supplies.

In short, the analysis indicated that benthic microfloral productivity was governed by a combination of variables that integrated into factors and were time

TABLE 9.7
List of Important Factors for the Ecological Data[37]

Important Factors	Variable Values	Isolated Effect on Mean (%)	Value of Productivity	Information Content on Adding Factor (%)
Sand				
1. Light	Low (750)	–59.6	14.6	52.48
Respiration	Low (10)			
Chlorophyll-a	Low (3)			
Tidal range	High (4)			
2. Light	High (1200)	47.3	53.0	93.27
Chlorophyll-a	High (5)			
Tidal range	Low (2)			
3. Light	High (1200)	–16.1	30.2	98.39
Chlorophyll-a	Low (3)			
Tidal range	Low (2)			
4. Light	High (1200)	–10.4	32.3	99.64
Respiration	Low (10)			
Chlorophyll-a	Low (3)			
Tidal range	High (4)			
Muddy Sand				
1. Light	Low (700)	–63.5	15.6	72.3
Chlorophyll-a	Low (6)			
Tidal range	High (4)			
Exposure	Low (4)			
2. Chlorophyll-a	High (10)	10.2	47.0	79.60
3. Light	High (1250)	26.5	53.9	87.64
Chlorophyll-a	High (10)			
Tidal range	Low (2)			
4. Light	Low (700)	29.7	55.3	96.30
Chlorophyll-a	High (10)			
Tidal range	High (4)			
Exposure	Low (4)			
5. Light	High (1250)	–16.1	35.8	99.59
Chlorophyll-a	Low (6)			
Tidal range	Low (2)			
Exposure	High (8)			

varying. The effects of the variables were stationary within factors, but varied across factors. Because most ecological data tend to behave in a similar fashion, the use of K-systems analysis is recommended for modeling ecological processes.

9.6 INTERACTION EFFECTS IN K-SYSTEMS

The main goal of K-systems analysis is to investigate the patterns that exist in the data without imposing an outside structure on it. This allows for the identification

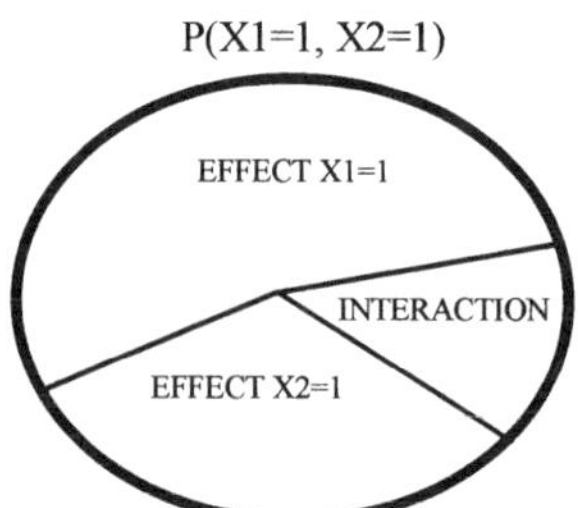

FIGURE 9.2 Effects of interactions.

of variables and the effects they have on the overall structure of the data. This is done by computing isolated variable effects and effects resulting from the interactions among variables. These are referred to as main and interaction effects, respectively. K-systems analysis computes the main and interaction effects using the principle of maximum entropy. In the analysis, solution properties are direct reflection of existing information.[10]

Figure 9.2 shows the classical view of effects and interactions. This view is fundamentally flawed. The dilemma arises from the fact that, while we can speak of the effect of $X_1 = 1$, once another variable is introduced into the system, the effect of $X_1 = 1$ is no longer a single effect, but is fractured into multiple effects. Given any set of marginal equations, the maximum entropy approximation to a higher-order distribution can be obtained. For example, we consider a function $P(X_1, X_2, X_3)$ where each X_i is either zero or one. The marginal equations are given as

$$^{1}P(X_1 = 1) = \tfrac{1}{3} = P(1,0,0) + P(1,1,0) + P(1,0,1) + P(1,1,1), \tag{9.17a}$$

and

$$^{2}P(X_2 = 1) = \tfrac{1}{4} = P(1,1,0) + P(1,1,1) + P(0,1,0) + P(0,1,1). \tag{9.17b}$$

Now, we add the probability normalization equation with each P initialized to $1/8$, $P(1,1,1) + P(1,1,0) + P(1,0,1) + P(1,0,0) + P(0,1,1) + P(0,1,0) + P(0,0,1) + P(0,0,0) = 1$, and iterate repeatedly through these three equations scaling the right-hand sides to satisfy the left-hand sides until convergence is achieved. Let $\mu(A,B)$ denote the maximum entropy approximation. It is important to note that the marginal sum $^{1}P(X_1 = 1) = P(1,0,0) + P(1,1,0) + P(1,0,1) + P(1,1,1)$ represents the effect, when considered relative to a flat system, or behavior attributable to $X_1 = 1$, and we can see the effect on $P(X_1, X_2, X_3)$ by computing $\mu(A)$. Similarly, the marginal sum $^{2}P(X_2 = 1) = P(1,1,0) + P(1,1,1) + P(0,1,0) + P(0,1,1)$ represents the effect, when considered relative to a flat system, or behavior attributable to $X_2 = 1$. Now, we can speak of the isolated effect due to $X_1 = 1$ or $X_2 = 1$. We put Equations 9.17a and b in the system together and compute $\mu(A,B)$ for

$$^{1}P(X_1 = 1) = P(1,0,0) + P(1,0,1) + P(1,1,1) + P(1,1,0), \tag{9.17a}$$

and $$^{2}P(X_2 = 1) = P(1,1,0) + P(1,1,1) + P(0,1,0) + P(0,1,1). \quad (9.17b)$$

We note that the different components will converge to different values. The effect of $X_1 = 1$ has been fractured into two effects which are represented by the small- and medium-sized P terms. The medium-sized component does not represent any form of interactions, and will converge under $\mu(A,B)$ to ${}^{1}P(X_1 = 1)\,{}^{2}P(X_2 = 1)$. That is, the medium-sized component treats X_1 and X_2 as independent variables. It no longer makes sense to speak of the effect of $X_1 = 1$ as a single component. Now, we add a third equation to the system. The new system is

$$^{1}P(X_1 = 1) = P(1,0,0) + P(1,0,1) + P(1,1,1) + P(1,1,0), \quad (9.17a)$$

$$^{2}P(X_2 = 1) = P(1,1,0) + P(1,1,1) + P(0,1,0) + P(0,1,1), \quad (9.17b)$$

and $$^{12}P(X_1 = 1, X_2 = 1) = P(1,1,0) + P(1,1,1). \quad (9.17c)$$

The difference in the medium-sized component for $\mu(A,B)$ and $\mu(A,B,C)$ can be attributed to the interaction of $X_1 = 1$ and $X_2 = 1$. Note that an interaction impacts not only the medium-sized component but also the large-size component. This enables us to speak of residual effects of an interaction.

Gouw[10-12] has extensively researched the interaction effects of K-systems. As stated earlier, an n-variable interaction in the context of K-systems theory represents an effect due to n variables acting in unison. It measures the combination effect as an increase in information content using information theory principles. A main effect is a special case of n-variable interaction. It is simply a one-variable interaction, that is, the interaction between the variable and the flat system. Gouw considered a 2^2 factorial experiment with and without interaction effects and concluded that when interaction effects are considerably large, the corresponding main effects have very little practical meaning.

9.6.1 Interaction Effects in a Plant Pathology Problem: A Case Study

This section is based on a case study conducted by Gouw[10] for a plant pathology design of experiments problem. The objective was to search for significant patterns in data and relationships among variables in the experiments. The data consisted of figures of dry weights of sorghum. These were obtained from a design of experiments conducted in two consecutive years. Because the analyses were similar for both years, we only present the analysis for the first year. A $3 \times 3 \times 2$ design of experiments with four repetitions was performed each year. Dry weights were categorized as root dry weights, stalk dry weights, and head dry weights.

TABLE 9.8
Details of Variables Under Plant Pathology Study[10]

Symbol	Variable	Level 1	Level 2	Level 3
M	*M. phaseolina*	0 cfu/g soil	10 cfu/g soil	100 cfu/g soil
N	Nematode	0 nematode	1554 nematodes/16 kg soil	3108 nematodes/16 kg soil
H	Hybrid	Dekalb–Pfizer 50	Pioneer 8333	—

The study consisted of the computation of percentages of variations and individual effects. The percentage of variation measured the variability in the data caused by deviation of data from its mean value. This helped measure the significance of main and interaction effects. After the significant factors (factors with a high percentage of variation) were known, the individual effects of these factors were studied.

The variables under investigation were *Macrophomina phaseolina*, nematode, and hybrid. *M. phaseolina* is a soil-borne fungus which destroys the root and stalk of sorghum. Nematode destroys root cells and reduces uptake of water and nutrients into the plant. Both severely infect plants preventing their growth as a result of root dysfunction.[34] The study investigated the effects of *M. phaseolina* and nematode for two different sorghum hybrids. Table 9.8 provides the details of the variables under this study, and Table 9.9 shows dry weights data with all possible combinations of levels of variables. Note that each cell in the table contains the four repetitions per level combination.[10]

9.6.2 Computation of Variations

The variations in data occur for a variety of reasons, such as differences in variables or external variables, measurement errors, or the data collection process. We can measure the importance of a variable by the proportion of total variation in the data. Illustratively, if two variables explain 81.5% and 5.5% of the total variation in the data, the second variable may be considered insignificant in most situations.

Let A, B, and C be three variables with their corresponding levels a, b, and c, respectively. Let α_i be the effect of A at level i, β_j be the effect of B at level j, γ_k be the effect of C at level k. Let $\alpha\beta_{ij}$ be the interaction between A and B at levels i and j; and $\alpha\beta\gamma_{ijk}$ be the interaction among A, B, and C at levels i, j, k. Let V_A, V_B, V_C, V_{AB}, V_{AC}, V_{BC}, and V_{ABC} be the variations explained by the variables (or variable combinations) A, B, C, AB, AC, BC, and ABC, respectively. Then,[33]

$$V_A = bc\sum_{i=1}^{a} \alpha_i^2, \tag{9.18a}$$

$$V_B = ac\sum_{j=1}^{b} \beta_j^2, \tag{9.18b}$$

TABLE 9.9
Dry Weights Data With All Possible Combinations of Levels of Variables[10]

Variable Levels			Root Dry Weight	Stalk Dry Weight	Head Dry Weight
M	N	H	(g)	(g)	(g)
1	1	1	(119, 98, 100, 102)	(56, 61, 54, 58)	(63, 50, 55, 54)
1	1	2	(121, 109, 130, 112)	(57, 62, 43, 50)	(55, 52, 56, 54)
1	2	1	(92, 94, 99, 93)	(47, 48, 44, 47)	(45, 40, 42, 46)
1	2	2	(92, 94, 93, 93)	(50, 44, 49, 47)	(31, 45, 40, 44)
1	3	1	(88, 91, 89, 90)	(49, 49, 46, 49)	(32, 40, 40, 43)
1	3	2	(89, 92, 84, 88)	(50, 42, 46, 46)	(30, 38, 42, 32)
2	1	1	(79, 90, 82, 77)	(46, 46, 44, 45)	(31, 34, 41, 26)
2	1	2	(78, 88, 76, 78)	(46, 46, 45, 46)	(15, 28, 24, 25)
2	2	1	(69, 80, 69, 70)	(45, 45, 44, 47)	(21, 29, 22, 22)
2	2	2	(70, 78, 63, 73)	(47, 47, 45, 45)	(32, 24, 23, 23)
2	3	1	(64, 70, 65, 67)	(42, 46, 40, 47)	(26, 26, 19, 20)
2	3	2	(64, 70, 64, 66)	(46, 40, 44, 45)	(24, 24, 20, 21)
3	1	1	(63, 66, 64, 65)	(47, 44, 46, 47)	(22, 21, 21, 20)
3	1	2	(64, 63, 63, 66)	(48, 45, 45, 44)	(23, 22, 22,22)
3	2	1	(66, 65, 64,64)	(44, 45, 46, 46)	(22, 21, 22, 23)
3	2	2	(63, 64, 65, 64)	(45, 44, 44, 45)	(20, 22, 25, 19)
3	3	1	(64, 66, 64, 64)	(46, 47, 40, 46)	(20, 21, 22, 22)
3	3	2	(62, 62, 63, 60)	(44, 46, 45, 44)	(20, 20, 23, 19)

$$V_C = ab\sum_{k=1}^{c}\gamma_k^2, \tag{9.18c}$$

$$V_{AB} = c\sum_{i=1}^{a}\sum_{j=1}^{b}\alpha\beta_{ij}^2, \tag{9.18d}$$

$$V_{AC} = b\sum_{i=1}^{a}\sum_{k=1}^{c}\alpha\gamma_{ik}^2, \tag{9.18e}$$

$$V_{BC} = a\sum_{j=1}^{b}\sum_{k=1}^{c}\beta\gamma_{jk}^2, \tag{9.18f}$$

and

$$V_{ABC} = \sum_{i=1}^{a}\sum_{j=1}^{b}\sum_{k=1}^{c}\alpha\beta\gamma_{ijk}^2. \tag{9.18g}$$

TABLE 9.10
Average Weights and Details of Significant Individual Main Effects and Interaction Effects for the Plant Pathology Data[10]

	Root Dry Weight		Stalk Dry Weight		Head Dry Weight	
Factor	Effect	% Average	Effect	% Average	Effect	% Average
Average	78.3	—	46.6	—	30.3	—
M1	19.7	25.2	3.1	6.7	14.2	46.9
M2	–5.4	–6.8	–1.7	–3.5	–5.3	–17.5
M3	–14.4	–18.3	–1.5	–3.2	–8.9	–29.4
N1	7.3	9.3	2.2	4.7	4.5	14.9
N2	–1.7	–2.2	–0.8	–1.7	–1.0	–3.4
N3	–5.5	–7.1	–1.4	–3.0	–3.5	–11.5
H1	–0.2	–0.2	0.3	0.7	0.9	3.0
H2	0.2	0.2	–0.3	–0.7	–0.9	–3.0
MN11	—	—	3.0	6.5	—	—
MN12	—	—	–1.9	–4.1	—	—
MN13	—	—	–1.1	–2.4	—	—
MN21	—	—	–1.6	–3.4	—	—
MN22	—	—	1.4	3.0	—	—
MN23	—	—	0.1	0.3	—	—
MN31	—	—	–1.5	–3.2	—	—
MN32	—	—	0.5	1.1	—	—
MN33	—	—	1.0	2.1	—	—

The total variation in the data, say V_{TOTAL}, is

$$V_{\text{TOTAL}} = V_A + V_B + V_C + V_{AB} + V_{AC} + V_{BC} + V_{ABC}. \tag{9.19}$$

This provides an important tool in measuring the importance of a factor. The factors which explain a high percentage of variation are considered significant. The factors which explain a low percentage of variation are considered less significant.

9.6.3 Data Analysis

K-systems analysis revealed that *M. phaseolina* was the most dominant factor in the data. It contributed at least 83.1%, 47.0%, and 84.7% to the variability in the root, stalk, and head dry weight data, respectively. Nematode explained for 11.5%, 23.3%, and 9.2% of all the variability in the root, stalk, and head dry weight data, respectively. Hybrid had very little influence on the outcome of experiments. Two- and three-variable interactions were negligible. However, two-variable interaction between *M. phaseolina* and nematode accounted for 23.4% of variability in the stalk dry weight data. Average weights and details of significant individual main effects and interaction effects for the data are given in Table 9.10.[10]

TABLE 9.11
An Example of a Nominal System

Humidity (v_1)	Leaf Loss (v_2)	Damage (f)
0	0	Mild
0	1	Terminal
1	0	Serious
1	1	Terminal

M. phaseolina and nematode interaction effects for stalk dry weight described an unusual relationship between *M. phaseolina* and nematode. The researcher concluded that the presence of either *M. phaseolina* or nematode in the soil would decrease the average dry weight of stalks. On the contrary, the presence of *M. phaseolina* and nematode in the soil, combinedly, would increase the average weight. Unlike the analysis of variance which introduces extraneous and, possibly, erroneous information in the data by forcing it to fit a particular model, K-systems analysis revealed important interaction effects.

9.7 CONCLUSION: TOWARD GENERAL SYSTEMS

We have emphased that RA and K-systems analysis have tremendous potential in modeling various natural and biological processes. It is evident that, with due care, the analysis will allow the most useful aspects of data to emerge in a clear and forceful way. However, there are situations where it is not possible to establish a direct relationship between the dependent and independent variables. For example, the dependent variable or system function may draw values from a nominal data set as is indicated in the example given by Table 9.11. We refer to such a system as a nominal system. One way to approach this problem is to regard a nominal function as a variable, thereby introducing one more variable to the set of variables, and simply observe the occurrence or nonoccurrence of the aggregate states. Theoretically, this transforms the original system into a selection system. Appropriate solution procedures can now be developed based on the methodology of RA and K-systems.

A selection system is a special case of a possibilistic system, where only occurrence or nonoccurrence of a state is recorded. On the other hand, it is possible to transform any arbitrary system into a selection system as stated above. From the perspective of learning behaviors of natural processes (i.e., discovering patterns in data), a selection system is simply a set of positive and negative examples. The occurrence of a state signifies a positive example, and the nonoccurrence of a state illustrates a negative example. Mathematically, we define a nominal system as the following six-tuple[5,44]

$$B_{no} = \left(V_{no}, W_{no}, s_{no}, A_{no}, Q_{no}, f_{no}\right) \tag{9.20}$$

where

1. $V_{no} = \{v_i \mid i \in \{1, 2, \ldots, n\}\}$ is a set of variables;
2. $W_{no} = \{V_j \mid j \in \{1, 2, \ldots, m\}, m \le n\}$ is a family of state sets;
3. $s_{no}: V_{no} \rightarrow W_{no}$ is an onto mapping which assigns to each variable in V_{no} one state set from W_{no};
4. $A_{no} = s_{no}(v_1) \times s_{no}(v_2) \times \cdots \times s_{no}(v_n)$ is a set of all potential aggregate states;
5. Q_{no} is a set of categories or nominal attribute values; and
6. $f_{no}: A_{no} \rightarrow Q_{no}$ is called the nominal system function which represents the information regarding the aggregate states of the system.

We transform the above system to a selection system as follows

$$B_{se} = \left(V_{se}, W_{se}, s_{se}, A_{se}, Q_{se}, f_{se}\right) \tag{9.21}$$

where

1. $V_{se} = \{v_i \mid i \in \{1, 2, \ldots, n + 1\}\}$ is a set of variables, variable v_{n+1} takes values from Q_{no};
2. $W_{se} = \{V_j \mid j \in \{1,2, \ldots, m\}, m \le n + 1\}$ is a family of state sets;
3. $s_{se}: V_{se} \rightarrow W_{se}$ is an onto mapping which assigns to each variable in V_{se}, one state set from W_{se};
4. $A_{se} = s_{se}(v_1) \times s_{se}(v_2) \times \ldots \times s_{se}(v_{n+1})$ is a set of all potential aggregate states;
5. $Q_{se} = \{0,1\}$; and
6. $f_{se}: A_{se} \rightarrow Q_{se}$ is a selection function which represents the information regarding the aggregate states of the system.

We define f_{se} as follows. For all aggregate states α_{no} and α'_{no} in B_{no},

$$f_{se}\left(\alpha_{no}\, f_{no}\left(\alpha'_{no}\right)\right) = \begin{cases} 1 & \text{if } \alpha_{no} = \alpha'_{no} \\ 0 & \text{otherwise.} \end{cases} \tag{9.22}$$

Table 9.12 represents an example of a transformed system using Equation 9.22. Apparently, there is no loss of information in transforming from one system to another. Thus, the new system must explain the behavior of the original system. K-systems analysis can now be invoked on this system. However, there are concerns while performing RA or K-systems analysis. An exhaustive search for factors is inherently exponential, which may limit the scope of analysis. Development of techniques is underway to counter such problems. Smart tools such as parallelization and genetic algorithms can be used in order to optimize the factor search.[8,15]

ACKNOWLEDGMENT

This work was partially supported by the Office of Naval Research grant number ONR N0014-90K-0611.

TABLE 9.12
An Example of the Transformed System of a Nominal System

Humidty, v_1	Leaf Loss, v_2	Damage, v_3	System Functio, f_{se}
0	0	Mild (0)	1
0	1	Mild (0)	0
1	0	Mild (0)	0
1	1	Mild (0)	0
0	0	Terminal (2)	0
0	1	Terminal (2)	1
1	0	Terminal (2)	0
1	1	Terminal (2)	1
0	0	Serious (1)	0
0	1	Serious (1)	0
1	0	Serious (1)	1
1	1	Serious (1)	0

REFERENCES

1. Ashby, W. R., Constraint analysis of many-dimensional relations, *Gen. Syst. Yearb.*, 9, 99–105, 1964.
2. Basar, E., *Biophysical and Physiological Systems Analysis*, Addison-Wesley, Reading, MA, 1976.
3. Cavallo, R. E. and Klir, G. J., Reconstructability analysis of multi-dimensional relations: a theoretical basis for computer-added determination of acceptable systems models, *Int. J. Gen. Syst.* 5, 143–171, 1979.
4. Cavallo, R. E. and Klir, G. J., Reconstructibility analysis: overview and bibliography, *Int. J. Gen. Syst.* 7, 1–6, 1981.
5. Cavallo, R. E. and Klir, G. J., Reconstructibility analysis: evaluation of reconstruction hypotheses, *Int. J. Gen. Syst.* 7, 7–31, 1981.
6. Cavallo, R. E. and Klir, G. J., Decision making in reconstructability analysis, *Int. J. Gen. Syst.*, 8, 243–255, 1982.
7. Cavallo, R. E. and Klir, G. J., Reconstruction of possibilistic behavior systems, *Fuzzy Sets Syst.*, 8, 175–197, 1982.
8. Dobransky, M. K. and Wierman, M. J., Genetic algorithm and reconstructability analysis, paper presented at the 2nd Workshop of the International Institute for General System Studies, January, San Marcos, TX, 1997.
9. Gatlin, L. L., *Information Theory and the Living System*, Columbia University Press, New York, 1972.
10. Gouw, D., Data Analysis Using Reconstructibility Theory, Ph.D. dissertation, Louisiana State University, Baton Rouge, May 1995.
11. Gouw, D. and Jones, B., Using K-systems theory to computer true main and interaction effects in a design of experiments, *Cybern. Syst.*, 26, 481–498, 1995.
12. Gouw, D. and Jones, B., The interaction concept of K-systems theory, *Int. J. Gen. Syst.*, 24, 163–169, 1996.
13. Higashi, M. and Klir, G. J., Measures of uncertainty and information based on possibility distributions, *Int. J. Gen. Syst.*, 9, 43–58, 1982.

14. Higashi, M., Klir, G. J., and Pittarelli, M. A., Reconstruction families of possibilistic structure systems, *Fuzzy Sets Syst.,* 12, 37–60, 1984.
15. Iles, P. E. and Jones, B., Parallelization of a greedy algorithm for a generalization of the reconstruction problem, *Adv. Syst. Sci. Appl.,* 215–220, 1995.
16. Jones, B., Determination of reconstruction families, *Int. J. Gen. Syst.,* 8, 225–228, 1982.
17. Jones, B., Recent algorithms in reconstructability analysis, *SGSR Proc.,* 243–245, 1984.
18. Jones, B., Determination of unbiased reconstruction, *Int. J. Gen. Syst.* 10, 169–176, 1985.
19. Jones, B., A greedy of algorithm for a generalization of the reconstruction problem, *Int. J. Gen. Syst.,* 11, 63–68, 1985.
20. Jones, B., The cognitive content of system substates, *Proceedings of the Symposium on the Cognitive Aspects of Reconstructability Analysis,* Palma de Mallorca, Spain, 1985.
21. Jones, B., Reconstructability analysis for general functions, *Int. J. Gen. Syst.,* 11, 133–142, 1985.
22. Jones, B., Reconstructability considerations with arbitrary data, *Int. J. Gen. Syst.* 11, 143–151, 1985.
23. Jones, B., K-systems versus classical multivariate systems, *Int. J. Gen. Syst.* 12, 1–6, 1986.
24. Jones, B., A program for reconstructability analysis, *Int. J. Gen. Syst.,* 15, 199–205, 1989.
25. Klir, G. J., Identification of generative structure in empirical data, *Int. J. Gen. Syst.,* 3, 89–104, 1976.
26. Klir, G. J. and Uyttenhove, H.J.J., Procedures for generating reconstruction hypotheses in the reconstructability analysis, *Int. J. Gen. Syst.,* 5, 231–246, 1979.
27. Klir, G. J., *Architecture of Systems Problem Solving,* Plenum Press, New York, 1985.
28. Klir, G. J. and Way, E.C., Reconstructability analysis: aims, result and open problems, *Syst. Res.,* 2(2), 141–163, 1985.
29. Klir, G. J., Reconstructability analysis: an offspring of Asby's constraint analysis, *Syst. Res.,* 3(4), 267–271, 1986.
30. Klir, G. J. and Parviz, B., Relationship between true and estimated possibilistic systems and their reconstructions, *Int. J. Gen. Syst.,* 12, 319–331, 1986.
31. Lendaris, G. G., Zwick, M., and Mathia, K., On matching ANN structure to problem domain structure, *Proceedings of World Congress on Neural Networks,* Portland, OR, July 1993.
32. Lewis, P. M., Approximating probability distributions to reduce storage requirements, *Inf. Control,* 2(3), 214–225, 1959.
33. Mason, R. L., Anderson, V. L., and Hess, J. L., Statistical Design and Analysis of Experiments, John Wiley and Sons, New York, 1989.
34. Mughogho, L. K. and Pande, J., Charcoal rot of sorghum, in *Sorghum Root and Stalk Rot: A Critical Review,* ICRISAT, Andhra Pradesh, India, 11–24, 1984.
35. Pittarelli, M., Uncertainty and estimation in reconstructability analysis, *Int. J. Gen. Syst.,* 15, 1–58, 1989.
36. Pittarelli, M., Reconstructability analysis using probability intervals, *Int. J. Gen. Syst.,* 16, 215–233, 1990.
37. Shaffer, G. P. and Cahoon, P., Extracting information from ecological data containing high spatial and temporal variability benthic microfloral production, *Int. J. Gen. Syst.,* 13, 107–123, 1987.

38. Shaffer, G. P. and Sullivan, M. J., Water column productivity attributable to displaced benthic diatoms in well mixed shallow estuaries, *J. Phycol.,* 24, 132–140, 1988.
39. Shaffer, G. P., K-systems analysis for determining the factors influencing benthic microfloral productivity in a Louisiana estuary, USA, *Mar. Ecol. Prog. Ser.,* 43, 43–54, 1988.
40. Shaffer, G. P., A comparison of benthic microfloral production on the west and gulf coasts of the United States: an introduction to the dynamic K-systems model, *Mar. Ecol. Prog. Ser.,* 43, 55–62, 1988.
41. Shannon, C. E., A mathematical theory of communications, *Bell Syst. Tech. J.,* 27(3), 379–423, July 1948.
42. Shore, J. E. and Johnson, R. W., Axiomatic derivation of the principle of maximum entropy and the principle of minimum cross entropy, *IEEE Trans. Inf. Theor.,* IT-26(1), 26–37, 1980.
43. Stent, G. S., *Molecular Biology of Bacterial Viruses,* W.H. Freeman, San Francisco, 1963.
44. Trivedi, S. K., Reconstructability Theory for General Systems and Its Applications for Automated Rule Learning, Ph.D. dissertation, Louisiana State University, Baton Rouge, Dec. 1993.
45. Trivedi, S. K. and Jones, B., Reconstructability of selection systems and U-structures, *Adv. Syst. Sci. Appl.,* 52–58, 1997.
46. Valdes-Perez, R. E. and Conant, R. C., Information loss due to quantization in reconstructability analysis, *Int. J. Gen. Syst.,* 9, 235–247, 1983.

10 Computer Simulation of Bimodal Neurons and Networks: Integrating Infrared and Visual Stimuli

Terry L. Huntsberger and John R. Rose

CONTENTS

10.1 INTRODUCTION

A reasonable starting point for investigating sensory fusion is an examination of systems that have successfully combined disparate types of sensory information. There are numerous biological neural network systems that integrate sensory information for

0-8493-7962-8/98/$0.00+$.50

an enhanced view of the environment. Among these are the barn owl,[29,30] the vestibulo-ocular reflex system,[22] the rattlesnake,[21] and the echolocating bats.[28] The optic tectum in the rattlesnake directly integrates bimodal inputs (visual and thermal infrared) in a set of six specialized neurons.[21] Our previous studies have indicated that these neuronal types can be used as generalized fusion filters for any pair of disparate sensory inputs.[11-13]

Aside from the fact that design features in this system may be directly applicable to the engineering problem of integrating visual and infrared (IR) stimuli, another reason for selecting this particular system is that it is more accessible than are other bimodal systems in more-advanced animals, such as vision and sonar in dolphins. The reptilian brain is significantly less complex than the mammalian brain.

This chapter employs large-scale computer simulations to investigate the characteristics of the unimodal and bimodal neurons in the optic tectum of the rattlesnake. The construction and simulation of large-scale neuronal networks using simplified models is governed by the trade-off between biological fidelity for a full model and the decrease in computational complexity for the simplified one. These concerns are present at the three modeling levels of somatic, dendritic/synaptic, and network interconnection. Reproduction of input/output relationships and behavior can be accomplished without the use of biologically plausible systems, such as the stereovision fusion-net system of Churchland[3] or the shape-from-shading system of Lehky and Sejnowski.[16,17]

On the other hand, temporal properties of biological neural networks are oftentimes difficult to incorporate in such models. This has been one of the motivations for the modeling of the CA3 region of the hippocampus using 9900 neurons by Traub and Miles,[32,33] the piriform cortex using 4600 cortical cells by Wilson and Bower,[1,34] and the frog cerebellar cortex using over 1.68 million cells by Pellionisz et al.[23] Different approaches were used in all of these studies for the characterization of the three modeling levels.

The next section discusses the various models that can be used depending on the level of detail desired in the simulation. This is followed by a description of the unimodal and bimodal neuron characteristics in isolation and in interconnected networks. A number of experimental studies are presented next. The chapter closes with a discussion of results and future directions of the research.

10.2 MULTILEVEL MODELING

10.2.1 Somatic Level

One of the earliest models derived from biological data for a single neuron was developed by Hodgkin and Huxley.[9] Although the model replicated the process of single-action potential generation, it was hampered by its inability to sustain subsequent activity. This property was partially addressed by a later study of Connor and Stevens.[4] At the extreme end of the pointlike neuron model are the logic units of McCulloch and Pitts.[20]

A compromise between the purely experimentally derived and logic unit methods can be achieved if models of potassium ionic transport, adaptive accommodative thresholds, and postfiring modulation of potassium conductance are used to generate a point soma neuron.[5,8,15,18] The governing equations for the point soma model[18] are:

$$\frac{dE}{dt} = \frac{-E + \left[P + G_K * \left(E_K - E\right)\right]}{\tau_{\text{LAT}}} \tag{10.1}$$

$$\frac{dT}{dt} = \frac{-\left(T - T_0\right) + c * E}{\tau_T} \tag{10.2}$$

$$S = \begin{cases} 0 & \text{if } E < T \\ 1 & \text{otherwise} \end{cases} \tag{10.3}$$

$$\frac{dG_K}{dt} = \frac{-G_K + B * S}{\tau_K} \tag{10.4}$$

where E is the transmembrane potential, P is the applied potential, G_K is the neuronal potassium conductance, E_K is the potassium equilibrium potential, τ_{LAT} is the latency time, T is the adaptive firing threshold, c is the threshold rise, T_0 is the resting threshold of the cell, τ_T is the time constant rise of the threshold, S is the spike generation parameter, B is the postfiring potassium conductance rise, and τ_K is the potassium decay time constant.

While somatic level modeling is adequate for simulating populations of unimodal neurons, it does not provide a mechanism for integrating disparate sensor information. Dendritic/somatic level modeling is required to address sensory fusion.

10.2.2 Dendritic/Synaptic Level

Decoupling the soma from its surrounding dendritic structure ignores the temporal and synaptic influences on neuronal activity.[27] In fact, Shepherd[26] has argued quite persuasively that the synapse is the primary functional unit of the nervous system. Inclusion of dendritic processes can be done at the chemical synaptic level using the quantitative models of Katz,[14] and/or the cable and compartmental models of Rall.[10,24,25] The recent work of Shepherd[27] and Wilson and Bower[34] has concentrated on optimizing the computational complexity of their models through reduced compartmental models.

The governing equations for active calcium-related conductances[18] are

$$\frac{dE}{dt} = \frac{-E + \left[P + G_{sd} * \left(V_{\text{Soma}} - E\right) + G_{\text{Ca}} * \left(E_{\text{Ca}} - E\right) + G_K * \left(E_K - E\right)\right]}{\tau_{\text{LAT}}} \tag{10.5}$$

$$\frac{dG_{\text{Ca}}}{dt} = \begin{cases} \dfrac{-G_{\text{Ca}}}{T_{G_{\text{Ca}}}} & \text{if } E < T \\ \dfrac{-G_{\text{Ca}} + D*(E-T)}{\tau_{\text{Ca}}} & \text{otherwise} \end{cases} \tag{10.6}$$

$$\frac{dC_{\text{Ca}}}{dt} = \frac{-C_{\text{Ca}} + A*G_{\text{Ca}}}{\tau_{\text{Ca}}} \tag{10.7}$$

$$\frac{dG_K}{dt} = \begin{cases} \dfrac{-G_K}{\tau_{G_K}} & \text{if } C_{\text{Ca}} < C_{\text{Ca}}0 \\ \dfrac{-G_K + S_K}{\tau G_K} & \text{otherwise} \end{cases} \tag{10.8}$$

where E is the transmembrane potential, P is the applied potential, G_{sd} is the conductance leakage from soma, V soma is the soma potential, G_{Ca} is the calcium conductance, E_{Ca} is th resting calcium potential, G_K is the neuronal potassium conductance, E_K is the potassium equilibrium potential, τ_{LAT} is the latency time, T is the adaptive firing threshold, c is the threshold rise, T_0 is the resting threshold of the cell, τ_T if the time constant rise of the threshold, S is the spike generation parameter, B is the postfiring potassium conductance rise, and τ_K is the potassium decay time constant.

Sensory fusion can be addressed by specifying different synaptic populations for the different sensory stimuli. Each synaptic population is then targeted by a separate fiber bundle.

10.2.3 Network Interconnection Level

The network that serves as the interconnection between neurons can be modeled at two different scales. These are[34]

- Small spatial subsample of actual network with realistic modeling of single neuron behavior. Long-range interactions would be difficult to model with this approach.
- Subsample of single neurons with coarser modeling of network dynamics. The handling of the missing cells can be accomplished through modification of synaptic interactions.

Both of these approaches are included in the GENESIS neural simulation system.[2] In addition to the spatial aspects of network modeling, there are temporal characteristics that also need to be considered.

10.2.4 Computational Concerns

Integration of the differential equations that govern the variables V, T, G_{Na}, G_{Ca}, and G_K are used for state updates of the individual neurons and the entire network. The solution technique must have the following properties:[19,34]

- Efficiency in order to run large network simulations,
- Accuracy in order to match results to experimental studies, and
- Stability in order to run simulations over relatively long time periods.

All of the state descriptions used throughout our simulation studies can be cast as first-order equations, which have a generic form of[18,34]

$$\frac{dx(t)}{dt} = -A(t)x(t) + B(t) \tag{10.9}$$

The solution to this first-order equation is

$$E_{t+\Delta t} = E_t e^{-A\Delta t} + \frac{B}{A}\left(1 - e^{-A\Delta t}\right) \tag{10.10}$$

where Δt should be chosen to be less than 1/5 of the fastest time constant that is a parameter in the model.[34]

For example, the equation for the membrane potential in the point soma model:

$$\frac{dE}{dt} = \frac{-E + \left[P + G_K * \left(E_K - E\right)\right]}{\tau_{\text{LAT}}} \tag{10.11}$$

can be rewritten as

$$\frac{dE}{dt} = -\left[\frac{1 + G_K}{\tau_{\text{LAT}}}\right] E + \left[\frac{P + G_K * E_K}{\tau_{\text{LAT}}}\right] \tag{10.12}$$

10.3 BIOLOGICAL SENSORY INTEGRATION

Snakes in the subfamily Crotalinae all possess pit organs that are in the front of the eyes as can be seen in Figure 10.1. The pit organs are sensitive to heat through a dense network of nerve fibers. These pit vipers also have specialized routing and processing centers, which control information transfer to the optic tectum portion of the midbrain. Although the optic nerve fibers feed almost directly into the optic tectum, only the trigeminal nerve from the pit organ passes through these centers, which are called the LTTD, and the reticularis caloris (RC). This process results in a type of spatial registration of the visual and thermal IR inputs within the optic tectum.

10.3.1 Unimodal Neurons

An extensive study of single thermal neuron properties in pit vipers was done by Bullock and Diecke (1956).[35] They observed that a low-intensity input caused a tonic firing, while firing became phasic as soon as a threshold was exceeded. The spatial response characteristics of thermal neurons was studied by Goris and Terashima.[6]

FIGURE 10.1 Timber rattlesnake showing thermal pit organs in front of eyes.

Two types of thermal neurons were found (cold and hot), which responded primarily as contrast detectors. This behavior is particularly important for discriminating prey that is either hotter or colder than the background, even while the snake is moving.

The spatiotopic mappings of both the thermal and visual portions of the optic tectum are relatively well matched in the "strike" zone in front of the rattlesnake.[7] A graphics recreation of the nonlinear mapping for the visual and thermal sensors derived from the data of Hartline et al.[7] is shown in Figures 10.2 and 10.3, respectively.

Most of the nonlinear portions of the maps are in the posterior region and in the peripheral portions of the thermal sensor. This indicates that processing in the LTTD prior to the optic tectum may provide some clues to the temporal properties of the rattlesnake response. The temporal properties of thermal neurons in the LTTD were recently measured by Terashima and Liang.[31] These parameters can be combined with the earlier work of Goris and Terashima[6] to give the operating characteristics of thermal neurons under different levels of stimulus.

10.3.2 Bimodal Neurons

Investigation of IR and visual integration in rattlesnakes provides evidence for several types of bimodal neurons. An early study indicated that bimodal neurons compromise approximately 10% of the units in layers 7a and 7b, 300 mm below the tectal surface.[7] The methodology employed indicated two types of bimodal units which they classified as OR and AND neurons. The OR neurons are characterized by their response to stimulation by either IR-only or visual-only signals or some combination of the two. In contrast, the AND neurons are characterized by their lack of response to stimulation by either IR-only or visual-only signals. The AND units respond only to stimulation consisting of a combination of visual and IR signals.

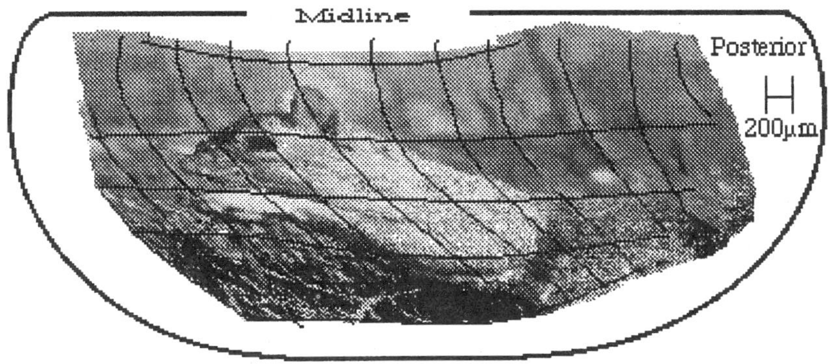

FIGURE 10.2 Visual spatiotopic map derived from Hartline et al.[7]

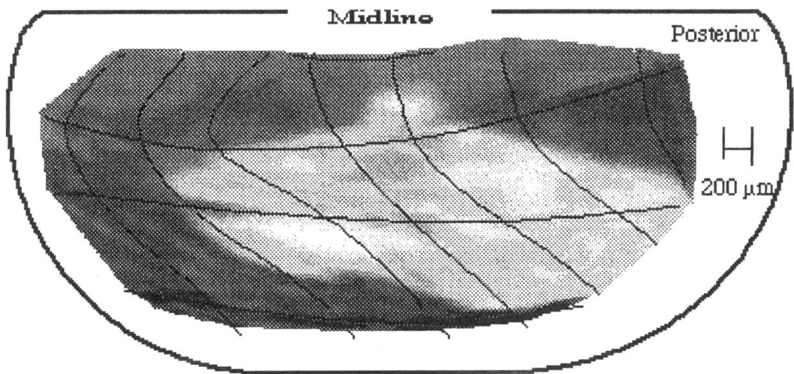

FIGURE 10.3 Infrared spatiotopic map derived from Hartline et al.[7]

A later study by Newman and Hartline,[21] employing unimodal-bimodal stimulus procedures and a rigorous comparison of response magnitudes, resulted in an adjusted estimate of the relative number of bimodal neurons. Newman and Hartline examined 196 tectal neurons, of which 103 showed some degree of cross-modality interaction. In their findings, they describe six types of bimodal neurons. In addition to the AND and OR units, the authors describe four types of units displaying cross-modality enhancement or depression. IR-enhanced visual and visual-enhanced IR neurons, when presented with a single stimulus modality, respond only to one of the two stimulus modalities. This stimulus is labeled the primary stimulus. The secondary stimulus, the other modality, does not elicit a response in the absence of the primary stimulus. However, the combination of primary and secondary stimuli results in a more forceful response than that which is brought about by the primary stimulus alone. In contrast, the secondary stimulus has the opposite effect in IR-depressed visual and visual-depressed IR neurons. The secondary stimulus causes an attenuation of response to the primary stimulus.

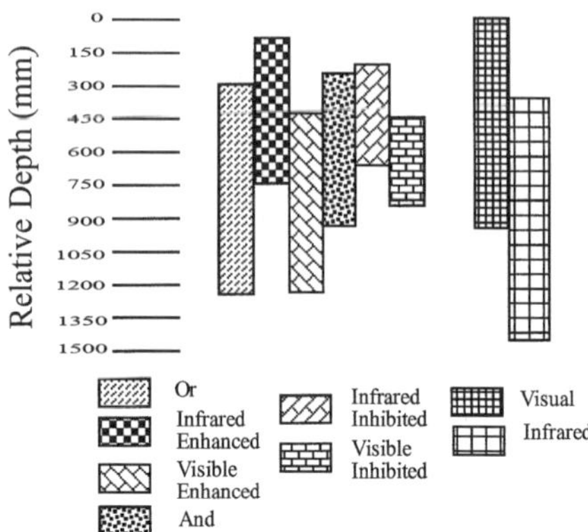

FIGURE 10.4 Relative depth map of neurons derived from Newman and Hartline.[21]

The relative depths of the six types of bimodal neurons and the thermal and visual neurons are shown in Figure 10.4. At depths of 450 to 750 mm there is a possibility of interactions between both unimodal and bimodal neurons. In this chapter we have chosen to concentrate on the simulation of unimodal neurons in isolation and in networks, and bimodal neurons in isolation. The next section presents some experimental studies along these lines.

10.4 EXPERIMENTAL STUDIES

For our simulation studies we decided to use a uniform set of algorithms such as those found in MacGregor.[18] This gave us a uniform approach to both the biological and computational aspects of the simulations. All references to the algorithms for specific portions of the experimental studies are to those in Part II.[18] Another system such as GENESIS[2] would have served as well.

10.4.1 Simulating Unimodal Neurons

In our model, spatiotopic neuronal behavior is approximated using a two-dimensional grid of point soma model neurons. Dendritic structure is captured in extracellular linkage of the outputs. The parameter settings for Equations 10.1 to 10.4 used in the first study are $E_K = 8.35$ mV for the potassium equilibrium potential, $c = 0.75$ for the threshold rise, $T_0 = -55.13$ mV for the resting threshold of the cell, $\tau_T = 1.76$ ms for the time constant rise of the threshold, $B = 20$ m·mho/cm^2 for the postfiring potassium conductance rise, and $\tau_K = 3.81$ ms for the potassium decay time constant. The values for these parameters used in our simulation are taken from the work of Terashima and Liang[31] and from MacGregor.[18]

The measured latency time for thermal neurons varied from 50 ms to no response in the studies of Goris and Terashima,[6] depending on the strength of the input stimulus. This means that a stimulus with a strong contrast to the background will cause rapid firing of neurons, with a highly nonlinear decrease in firing frequency

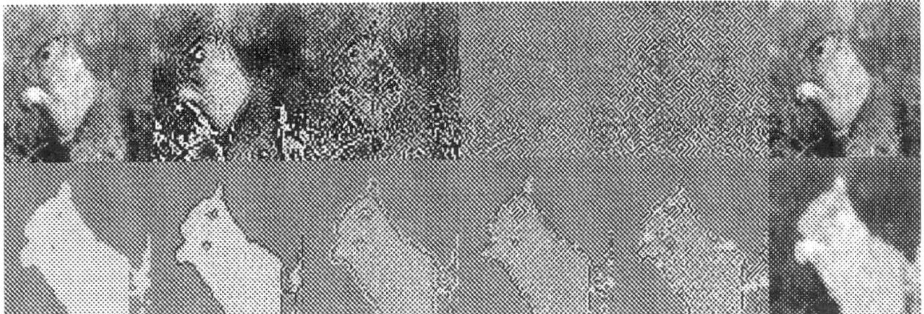

FIGURE 10.5 Response of two-dimensional array of synthesized neurons in steps of 1 ms; (top) visual; (bottom) thermal; original input shown on the far right of each row.

as the contrast decreases. This effect can be included in the model using an exponential weighting of τ_{LAT} of the form

$$\tau'_{LAT} = \tau_{LAT} e^{(1.0-P)/2.0} \quad (10.13)$$

for normalized input stimuli P between 0.5 and 1.0, and

$$\tau'_{LAT} = \tau_{LAT} e^{(1.0-P)/1.05} \quad (10.14)$$

for normalized stimuli P less than 0.5, where the latency time measured with direct electrical stimulation is given by τ_{LAT}. We modeled the different behavior of visible and thermal neurons by using τ_{LAT} of 0.8 ms for the visual neurons and the modified form of τ'_{LAT} for the thermal neurons.

Rapid accommodation to a static input scene occurs in the rattlesnake, and the AND fusion neurons usually only respond to rapidly moving stimuli.[7] This response is of use to the rattlesnake to filter out background clutter, since prey will usually be moving. Figure 10.5 shows a temporal study of the visual and thermal neuron arrays excited by a static stimulus performed using a modified version of the POOL20 algorithm.[18] The time step between frames is 1 ms and the array resolution is 64 × 64 neurons. Gray scale in this series represents the transmembrane potential, with dark for negative values and light for positive values. Although there is some enhancement of the edges in the visual array, by the third frame it has accommodated to the pattern and feature discrimination is barely possible. The thermal array immediately performs a contrast enhancement of the input due to the nonlinear latency behavior, and continues to favor the higher thermal area throughout the sequence.

This simulation assumes that the extracellular linkage is only characterized by a time delay, and there are no interactions between neurons other than nearest neighbor. In addition, there was no interaction between the visual and thermal neurons. The actual system in the rattlesnake is fed by fibers from the LTTD, and excitatory and inhibitory interactions between neurons will have different temporal properties for visual and thermal neurons. The next section describes a study that includes these effects.

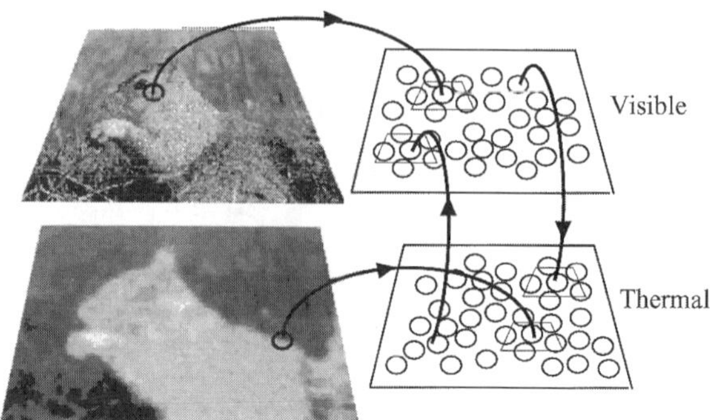

FIGURE 10.6 Interacting two-dimensional arrays of visual and thermal neurons, where fiber bundles are used to feed small neighborhoods in each array.

10.4.2 Simulating Interactions between Unimodal Neurons

Including the temporal characteristics of neuronal excitation into a simulation can be done using the NTWRK22 algorithm.[18] Two arrays of size 128 × 64 were directly fed through fibers from the visual and thermal sensors. We used the mappings of Hartline et al.[7] shown in Figures 10.2 and 10.3 in order to simulate the alignment seen in the optic tectum. There are fiber bundles running in both directions between the two arrays for either excitatory or inhibitory interactions. This system is shown in Figure 10.6, where the fiber bundles are shown for selected neurons in the arrays. The arrows on the links between the arrays show the direction of feed, with a neighborhood of interaction on the receiving end. The neighborhood size was 5 × 5 for the visual array and 9 × 9 for the thermal array. We used different size neighborhoods in order to simulate the relative differences in spatial sensitivity of the sensors.

Unlike the simulation in the previous section, the feed from the sensors is based on a probabilistic process. This means that all of the sensor neurons will not be firing at the same time, and there will be a nonuniform response in each of the visual and thermal arrays. The response of the thermal array for time sampled every 5 ms is shown in the left column of Figure 10.7, where gray scale is once again being used to represent the transmembrane potential. There is some activity in the area where the squirrel is, but compared with Figure 10.5, there is virtually no coordinated response.

Since the visual and thermal neurons are spatially coincident as seen in Figure 10.4, there will be interaction between them. This interaction can be simulated by turning on the fibers between the arrays (see Figure 10.6). For an excitatory interaction, the response of the visual array should be stronger, as is seen in thermally enhanced visual bimodal cells. This effect can be seen in the right column of Figure 10.7, where there is sustained activity in the thermally "hot" area of the squirrel.

10.4.3 Simulating Bimodal Neurons

Although Newman and Hartline[21] report six types of bimodal neurons, they also indicate that more than half of the neurons that they examined exhibited some degree

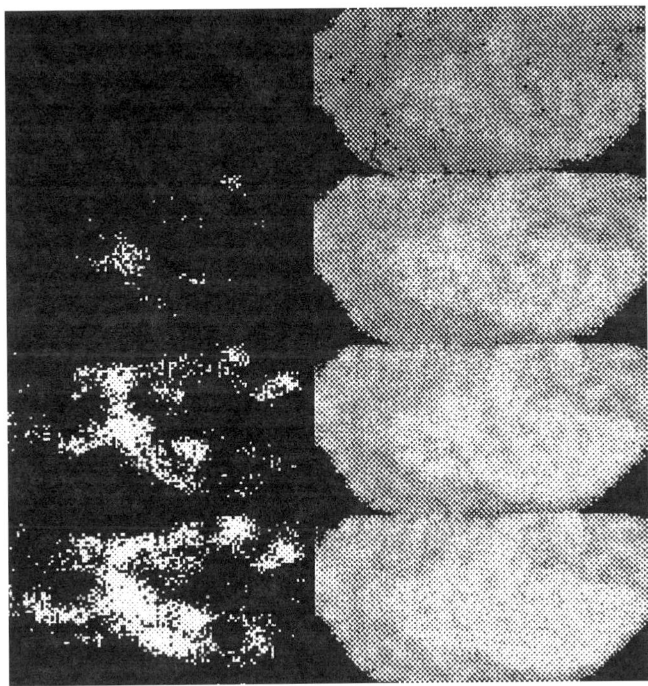

FIGURE 10.7 Response of two-dimensional array of synthesized neurons sampled every 5 ms; (left) visual array with no thermal interaction; (right) visual array with thermal enhancement.

of cross-modal interaction. In addition, they report that several units shared properties of two or more types. This suggests that it is unlikely that six fundamentally different mechanisms are required to explain the various types of bimodal units. The hypothesis underlying our simulation of bimodal neurons is that variations of two underlying mechanisms suffice to reproduce the six types of cross-modal interaction that Newman and Hartline report. We are able to simulate AND, OR, IR-enhanced visual, and visual-enhanced IR units by making adjustments to a single model. We use a second related model in order to simulate IR-depressed visual and visual-depressed IR units. Both of these simple models are based on MacGregor's neuron model DENDR51.[18] This model was chosen because it allows one to specify details of the dendritic process. In particular: the morphology of the dendritic tree is user specified. Active calcium-related conductances are built in. The model includes the representation of synaptic action by equivalent synaptic conductances representing the influence of dendritic spines. An arbitrary number of synaptic input systems, with individually adjustable synaptic action, can be specified. Also, the model allows the user considerable control over the way in which synaptic systems are targeted.

Both models that we developed use the dendritic tree shown in Figure 10.8. This tree consists of 30 branches of equal length. Each branch is in turn broken into ten compartments. The models incorporate two populations of fibers. Each fiber population is associated with a particular stimulus modality, IR or visual. In the case of the model used to simulate OR, AND, IR-enhanced visual, and visual-enhanced IR

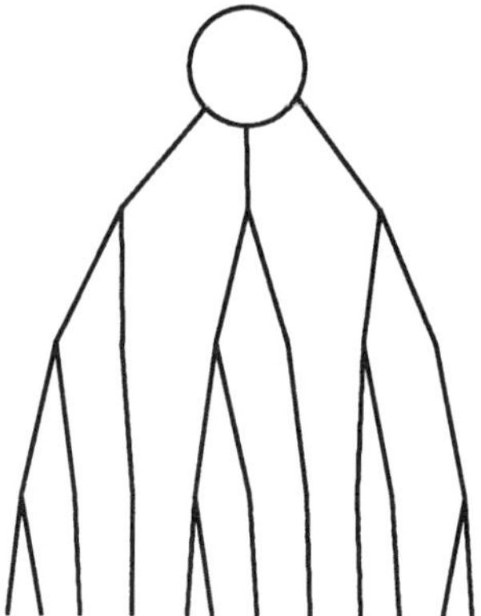

FIGURE 10.8 The dendritic tree used in simulating bimodal neurons.

units, the fibers target two similar populations of synapses. These synaptic populations have the same distribution, but their densities vary depending on which type of bimodal neuron is being simulated. In contrast, the two fiber populations in the model used to simulate IR-depressed visual and visual-depressed IR neurons target two distinct populations of synapses. One population of synapses is excitatory. The excitatory synapses are targeted by the fiber population corresponding to the primary stimulus. The other synapse population is inhibitory and is targeted by the fiber population corresponding to the second stimulus.

10.4.3.1 Simulating OR Units

The group of bimodal neurons classified as OR neurons may be stimulated by an IR-only signal or by a visual-only signal. Interestingly, it has been reported that these neurons may be stimulated more vigorously by combined submaximal visual and IR stimulation than by either one alone. A possible explanation for the mechanism underlying the observed response of the OR neurons is that the dendritic structure receives synaptic bombardment from efferent fibers from visual and IR neurons of sufficient strength so that signals from either source are adequate to elicit a response. This hypothesis was tested by simulating an OR neuron using MacGregor's neuron model DENDR51. The definition of the model entailed specifying two fiber populations corresponding to the efferent fibers from visual and IR neurons. In light of the lack of experimental evidence to the contrary, the assumption was made that both fiber populations involved the same synaptic type. The model was designed so that stimulation from either of the fiber populations would elicit a response. Consequently, the two synaptic populations have the same distribution and density. Furthermore, they are targeted by the same number of fibers. Figure 10.9 shows the response of the model corresponding to moderate stimulation from one fiber population. All simulations were run over 180 ms. At the beginning of each run, current is injected into the soma so that it stabilizes at −64 mV. Synaptic bombardment of the neuron dendritic structure commences 90 ms into the simulation and terminates 140 ms into the simulation. Since both fiber populations were defined

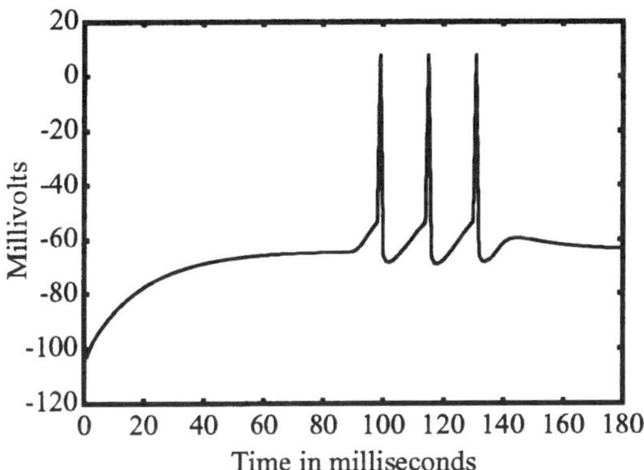

FIGURE 10.9 Simulation of a bimodal OR neuron with input from only one fiber population.

with the same characteristics, the results shown in Figure 10.9 are the same for stimulations corresponding to either a visual or IR source.

The first panel of Figure 10.10 shows the response of the model to submaximal stimulation from only one source. As mentioned previously, the same result is obtained regardless of whether the source of the stimulation is the fiber population–simulating visual input or the fiber population–simulating IR input. In contrast, the second panel of Figure 10.10 shows the response of the model to the combined submaximal visual and IR stimulation. Here the submaximal stimulation that appeared in only one fiber population in the first panel appears in both fiber populations. This is consistent with reported experimental results which indicate some intermodality summation.[7]

10.4.3.2 Simulating Direction-Specific OR Units

Hartline et al.[7] have also observed direction-specific OR neurons. These neurons appear to respond maximally to an IR or visual stimulus in one direction but do not respond to the same stimulus in the opposite direction. Suboptimal responses are observed when the stimulus is presented along some other axis. Such neurons are critical for the detection and integration of motion information and for tracking. Simulating such neurons necessarily entails a more complex model since it is possible that there is an interaction of excitatory as well as inhibitory elements that must be included. We have been able to produce a first approximation of this behavior by controlling the order in which various dendritic regions are stimulated. Controlling the order in which regions are stimulated makes it possible to specify a moving wave of stimulation across the dendritic structure in a given direction. This entailed redesigning the portion of the model in which the signals from input fibers are updated. This portion was first modified so that the stimulation of the dendritic tree would proceed in a distal-to-proximal direction in alignment with the dendritic structure. The left-hand panel of Figure 10.11 shows the result of this simulation. The direction-specific OR unit responds to stimulation in this direction. Next, the model was modified so that the stimulation would proceed in a proximal-to-distal

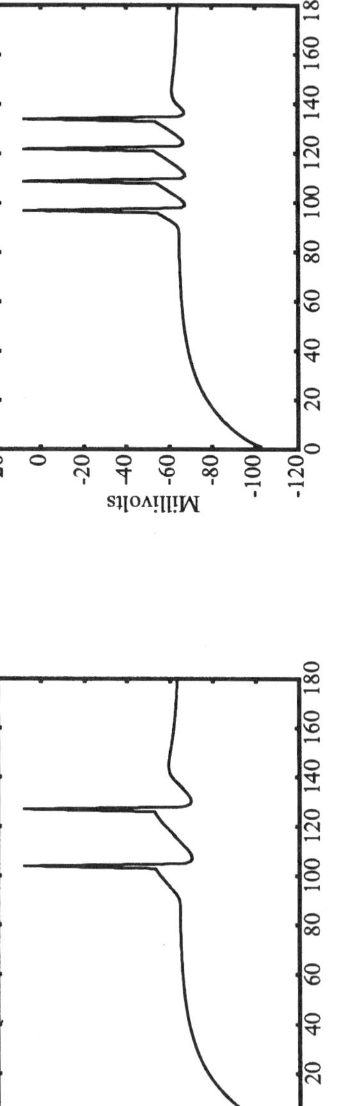

FIGURE 10.10 The first panel shows the simulation of a bimodal OR neuron with submaximal input from only one fiber population. The second panel shows the simulation of a bimodal OR neuron with submaximal input from both fiber populations.

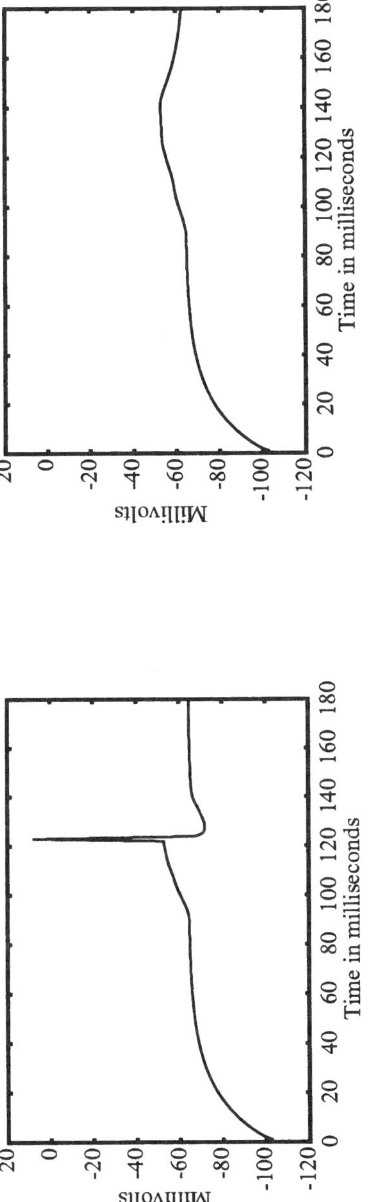

FIGURE 10.11 The left-hand panel shows the result of simulating a direction-specific OR neuron in which the stimulus proceeds in a distal-to-proximal direction in alignment with the dendritic structure. The right-hand panel shows the result of simulating a direction-specific OR neuron in which the stimulus proceeds in a proximal-to-distal direction in alignment with the dendritic structure.

direction in alignment with the dendritic structure. The right-hand panel of Figure 10.11 shows the result of stimulation in this direction. The unit does not respond to this stimulation even though the stimulation is identical in all respects to that used in the right-hand panel except for the direction of motion.

10.4.4.3 Simulating AND Units

Another class of neurons is stimulated only by a combination of visual and IR signals and has been classified as AND neurons. These neurons have been described as responding to only the first few iterations of a stimulus and as often requiring rapid stimulus motion. Our explanation of the underlying mechanism of these neurons builds on our hypothesized mechanism for the OR neuron. Evidence suggests that these neurons require stimulation from a substantial percentage of their synaptic connections from efferent fibers from visual and IR neurons in order to fire. If this is the case, then one would expect that as some of the neurons stimulating the bimodal AND neuron accommodate, the level of stimulation would drop below the threshold required to elicit a response. Consequently, stimulation from only one fiber population is unlikely to result in the neuron being triggered. In fact, it might be the case that bimodal AND neurons are simply OR neurons with limited synaptic connections. This hypothesis was tested by simulating an AND neuron using the same model for the OR neuron with changes to the density of the two synaptic populations. The definition of the model entailed specifying two fiber populations corresponding to the efferent fibers from visual and IR neurons. In light of the lack of experimental evidence to the contrary, the assumption was made that both fiber populations involved the same synaptic type as in the case of the simulation of the OR neuron. The model was designed with limited synaptic connections to the two fiber populations so that stimulation from both fiber populations would be required to elicit a response.

The first panel of Figure 10.12 shows the response of the model corresponding to stimulation from one fiber population. Since both fiber populations were defined with the same characteristics, the result shown in the first panel of Figure 10.12 are the same for stimulations corresponding to either a visual or IR source. In this figure we see that the stimulation is not sufficient to cause the AND neuron to fire.

In contrast, the second panel of Figure 10.12 shows the response of the model to stimulation from both populations of fibers, corresponding to combined visual and IR stimulation. Here the stimulation that appeared in only one fiber population in the right-hand appears in both fiber populations and is sufficient to elicit a response from the AND neuron.

10.4.3.4 Simulating Enhanced Units

Two types of enhanced bimodal neurons have been found in the optic tectum of the rattlesnake. IR-enhanced visual neurons and visual-enhanced IR neurons respond to their primary stimulus, visual and IR, respectively. However, stimulation by their second stimulus modality, in the absence of a primary stimulus fails to elicit a response. Presenting both primary and secondary stimuli simultaneously produces

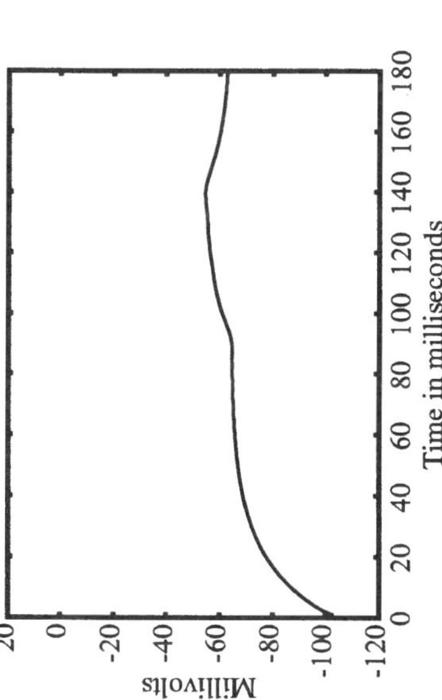

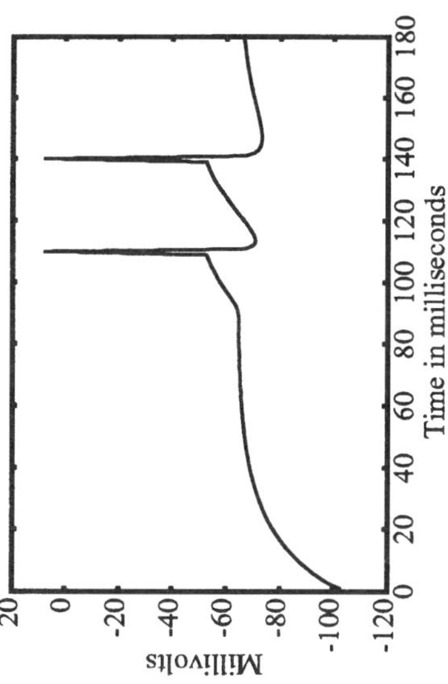

FIGURE 10.12 The left-hand panel shows the simulation of a bimodal AND neuron with input from only one fiber population. The right-hand panel shows the simulation of a bimodal AND neuron with input from both fiber populations.

an enhanced response. We use the same model to simulate IR-enhanced visual and visual-enhanced IR neurons. This model combines features from the models used to simulate OR and AND neurons. In particular, the secondary stimulus is modeled the same way the stimuli to the AND neuron is modeled so that it does not elicit a response on its own. The left-hand panel in Figure 10.13 shows the response of the neuron to stimulus by the secondary modality. In contrast, the primary stimulus is modeled the same way the stimuli to the OR neuron are modeled so that it is able to cause the neuron to fire. The right-hand panel of Figure 10.13 shows the response when the neuron is stimulated by its primary stimulus modality. Combining both inputs results in a higher level of stimulation since they are integrated. This is shown in Figure 10.14.

10.4.3.5 Simulating Depressed Units

The optic tectum of the rattlesnake also includes two types of depressed bimodal neurons. IR-depressed visual neurons and visual-depressed IR neurons respond to their primary stimulus in the absence of the secondary stimulus. Presenting the primary and secondary stimuli simultaneously results in a depressed response.

The model used to simulate depressed neurons differs fundamentally from the underlying model that we used to simulate the other types of bimodal neurons in the rattlesnake optic tectum. Unlike that model, this model requires that one population of synapses be inhibitory. Consequently, the population of synapses targeted by fibers from the secondary stimulus have a depressing effect on the neuron. The other populations of synapses, those targeted by fibers from the primary stimulus, are the same as the synapse populations used in the other models. The left-hand panel of Figure 10.15 shows the response of the depressed neuron to stimulation by the primary stimulus modality. In the absence of the secondary stimulus, this elicits a response. The right-hand panel shows the response of the neuron to the secondary modality. Stimulation by the secondary modality has a depressing effect on the soma potential.

Figure 10.16 shows the response of the neuron to simultaneous stimulation by both primary and secondary stimuli. The secondary stimulus depresses the neuron sufficiently to prevent the neuron from triggering. Newman and Hartline[21] report that depression in these two types of units varies considerably. At one end of the spectrum, response to primary stimulation was abolished completely by simultaneous presentation of primary and secondary stimuli as shown in Figure 10.16. At the other end of the spectrum, the primary response was only depressed by 5 to 10%. They go on to generalize that the weaker the primary response, the greater the degree of depression produced by secondary stimulation.

10.5 SUMMARY AND CONCLUSIONS

This chapter has presented simulation studies of the unimodal and bimodal neurons found in the optic tectum of the rattlesnake. The rattlesnake has a unique combination of visual and thermal senses feeding into neurons that directly integrate their inputs. We have included both spatial and temporal effects taken from the experimental

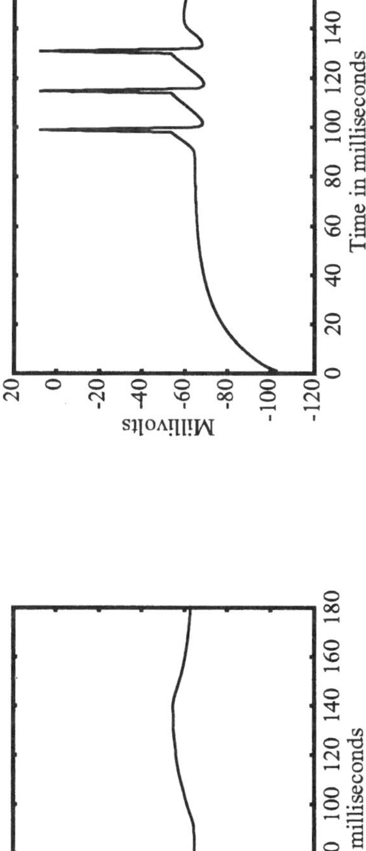

FIGURE 10.13 The left-hand panel shows the response of the enhanced bimodal neuron to stimulation by the secondary stimulus modality. The right-hand panel shows the response of the neuron to stimulation by the primary stimulus modality.

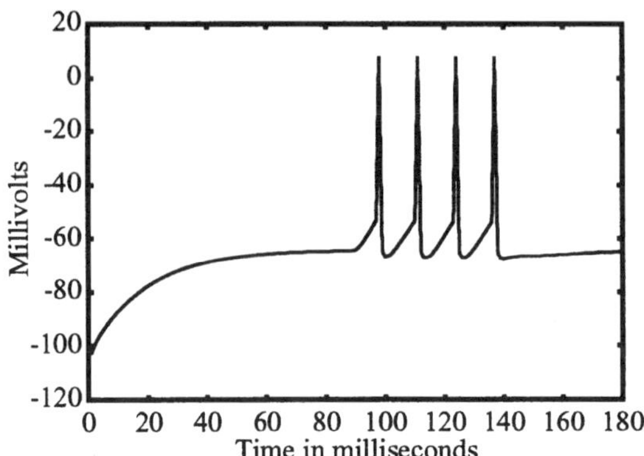

FIGURE 10.14 Stimulation of an enhanced bimodel neuron by both primary and secondary stimulus modalities.

literature into our simulations. Our experimental simulation studies indicate that the unimodal thermal neurons in isolation tend to act as a contrast enhancement mechanism due to nonlinear latency properties, whereas the visual neurons quickly accommodate to the input signal. When the visual and thermal neurons are allowed to interact with each other, this contrast enhancement behavior tends to enhance the coordinated firing of neurons in the visual neurons which are located in the thermally active regions of the scene.

The results of our simulations of the six classes of bimodal neurons reported by Newman and Hartline are consistent with experimental observations. The OR neurons are modeled using two synaptic populations with similar properties and similar distribution densities. Direction-specific OR neurons are modeled by controlling the order in which dendritic regions are stimulated. We model the AND neuron as a type of OR neuron with marginal synaptic connection from visual and IR fibers so that simultaneous stimulation from both populations of fibers is required in order for the AND neuron to be triggered. We model enhanced neurons by combining features from the models for AND and OR models. In particular, the synapse population associated with the primary stimulus is modeled after our OR model and the synapse population associated with the secondary stimulus modality is modeled after the AND model. In the case of depressed bimodal neurons, we model the secondary stimulus which depresses the neuron by using an inhibitory synaptic population.

By making small adjustments to the synaptic populations we are able to simulate six types of bimodal neurons using only two basic models. Although the results of our simulations are consistent with experimental observations, this alone does not prove that our models are correct. However, these results in conjunction with the economy and simplicity of our models would tend to indicate that a more-involved set of models is not necessary in order to explain the observed behavior of these neurons.

We are currently extending our results to include dynamic sequences of input for the visual and thermal senses. In addition, more complete studies of the direction

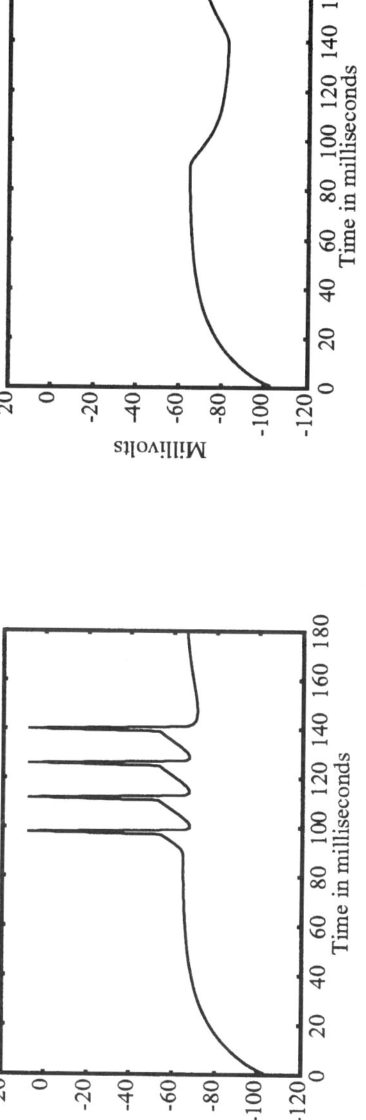

FIGURE 10.15 The left-hand panel shows the response of a depressed bimodal unit to its primary stimulus. The right-hand panel shows the response of the same unit to its secondary stimulus.

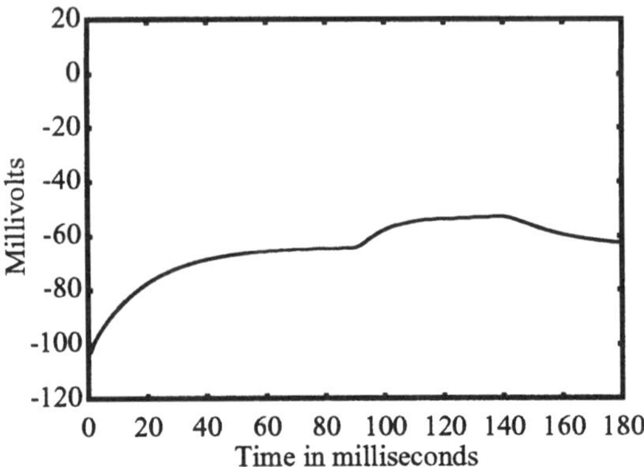

FIGURE 10.16 The attenuated response of a depressed bimodal unit to the simultaneous presentation of primary and secondary stimuli.

characteristics of the dendritic tree structure are currently being done. Inclusion of a full complement of unimodal and bimodal populations will give a more realistic simulation of the actual optic tectum. A compilation of the neuronal behavior parameters into tables can be used to address the engineering problem of sensory fusion for large, real-time visual and thermal inputs.

ACKNOWLEDGMENTS

The first author was supported in part under ONR Grant No. N00014-94-1-1163 and ARO Grant No. DAAH04-96-10326. We would like to thank Bev Huntsberger for generating the spatiotopic mapping figures.

REFERENCES

1. Bower, J. M., Relations between the dynamical properties of single cells and their networks in piriform (olfactory) cortex, in *Single Neuron Computation,* Zornetzer, S. F., McKenna, T., and Davis, J., Ed., Academic Press, Boston, 1992, 437–462.
2. Bower, J. M. and Beeman, D., *The Book of GENESIS,* Springer-Verlag/TELOS, Santa Clara, CA, 1995.
3. Churchland, P. M., A feed-forward network for fast stereo vision with a movable fusion plane, in *Android Epistemology: Proc. 2nd Int. Workshop on Human and Machine Cognition,* Ford, K. and Glymour, C., Eds., AAAI Press/MIT Press, Cambridge, MA, 1992.
4. Connor, J. A. and Stevens, C. F., Prediction of repetitive firing behavior from voltage clamp data on an isolated neuron soma, *J. Physiol.* (London), 213, 31–54, 1971.
5. Eccles, J. C., *The Physiology of Nerve Cells,* Academic Press, New York, 1957.
6. Goris, R. C. and Terashima, S., Central response to infra-red stimulation of the pit receptors in a Crotaline snake, *Trimeresurus flavoviridis, J. Exp. Biol.,* 58, 59–76, 1973.
7. Hartline, P. H., Kass, L., and Loop, M. S., Merging of modalities in the optic tectum: infrared and visual integration in rattlesnakes, *Science,* 199, 1225–1229, 1978.

8. Hill, A. V., Excitation and accommodation in nerve, *Proc. R. Soc. London Ser. B,* 119, 305–355, 1936.
9. Hodgkin, A. L. and Huxley, A. F., A quantitative description of membrane current and its application to conduction and excitation in nerve. *J. Physiol.* (London), 117, 500–544, 1952.
10. Holmes, W. R. and Rall, W., Electrotonic models of neuronal dendrites and single neuron computation, in *Single Neuron Computation,* Zornetzer, S. F., McKenna, T., and Davis, J., Ed., Academic Press, Boston, 1992, 7–26.
11. Huntsberger, T. L., Comparison of techniques for disparate sensor fusion, in *Proc. SPIE Symp. on Sensor Fusion III: 3-D Perception and Recognition,* Vol. 1383, Boston, November 1990, 589–594.
12. Huntsberger, T. L., Data fusion: a neural networks implementation, in *Data Fusion in Robotics and Machine Intelligence,* Abidi, M. A. and Gonzalez, R. C., Eds., Academic Press, Orlando, FL, 1992.
13. Huntsberger, T. L., Sensor fusion in a dynamic environment, in *Proc. SPIE Symposium on Sensor Fusion V,* Schenker, P., Ed., Vol. 1828, Boston, November 1992, 175–182.
14. Katz, B., *The Release of Neural Transmsitter Substances,* Charles C. Thomas, Springfield, IL, 1969.
15. Kernell, D., The repetitive impulse discharge of a simple neuron model compared to that of spinal motoneurons, *Brain Res.,* 11, 685–687, 1968.
16. Lehky, S. R. and Sejnowski, T. J., Network model for shape-from-shading: neural function arises from both receptive and projective fields, *Nature* (London), 333, 452–454, 1988.
17. Lehky, S. R. and Sejnowski, T. J., Neural network model of visual cortex for determining surface curvature from images of shaded surfaces, *Proc. R. Soc. Lond. B,* 240, 251–278, 1990.
18. MacGregor, R. J., *Neural and Brain Modeling,* Academic Press, San Diego, CA, 1987.
19. Mascagni, M. V., Numerical methods for neuronal modeling, in *Methods in Neuronal Modeling,* Koch, C. and Segev, I., Eds., MIT Press, Cambridge, MA, 1992, 439–484.
20. McCulloch, W. S. and Pitts, W., A logical calculus of the ideas imminent in nervous activity, *Bull. Math. Biophys.,* 5, 115–133, 1943.
21. Newman, E. A. and Hartline, P. H., Integration of visual and infrared information in bimodal neurons of the rattlesnake tectum, *Science,* 213, 789–791, 1981.
22. Paulin, M. G., Nelson, M. E., and Bower, J. M., Neural control of sensory acquisition: the vestibulo-ocular reflex, in *Advances in Neural Information Processing Systems,* Vol. I, Touretzky, D. S., Ed., Morgan-Kaufmann Publishers, San Francisco, 1989, 410–418.
23. Pellionisz, A., Llinas, R., and Perkel, D. H., A computer model of the cerebellar cortex of the frog, *Neuroscience,* 2, 19–35, 1977.
24. Rall, W., Time constants and electrotonic length of membrane cylinders and neurons, *Biophys. J.,* 9, 1483–1508, 1969.
25. Rall, W., Core conductor theory and cable properties of neurons, in *Handbook of Physiology (Sect. 1): The Nervous System I. Cellular Biology of Neurons,* Kandel, E. R., Ed., American Physiological Society, Baltimore, MD, 1977, 39–97.
26. Shepherd, G. M., *The Synaptic Organization of the Brain,* 2nd ed., Oxford University Press, New York, 1979.
27. Shepherd, G. M., Canonical neurons and their computational organization, in *Single Neuron Computation,* Zornetzer, S. F., McKenna, T., and Davis, J., Eds., Academic Press, Boston, 1992, 27-60.

28. Simmons, J. A., Acoustic-imaging computations by echolocating bats: unification of diversely-represented stimulus features into whole images, in *Advances in Neural Information Processing Systems,* Vol. II, Touretzky, D. S., Ed., Morgan-Kaufmann Publishers, San Francisco, 1990, 2–9.
29. Spence, C. D. and Pearson, J. C., The computation of sound source elevation in the barn owl, in *Advances in Neural Information Processing Systems,* Vol. II, Touretzky, D. S., Ed., Morgan-Kaufmann Publishers, San Francisco, 1990, 10–17.
30. Spence, C. D., Pearson, J. C., Gelfand, J. J., Peterson, R. M., and Sullivan, W. E., Neuronal maps for sensory-motor control in the barn owl, in *Advances in Neural Information Processing Systems,* Vol. I, Touretzky, D. S., Ed., Morgan-Kaufmann Publishers, San Francisco, 1989, 366–3734.
31. Terashima, S. and Liang, Y., Temperature neurons in the Crotaline trigeminal ganglia, *J. Neurophysiol.,* 66, 623–634, 1991.
32. Traub, R. D. and Miles, R., *Neuronal Networks of the Hippocampus,* Cambridge University Press, New York, 1991.
33. Traub, R. D. and Miles, R., Synchronized multiple bursts in the hippocampus: a neuronal population oscillation uninterpretable without accurate cellular membrane kinetics, in *Single Neuron Computation,* Zornetzer, S. F., McKenna, T., and Davis, J., Ed., Academic Press, Boston, 1992, 463–476.
34. Wilson, M. A. and Bower, J. M., The simulation of large-scale neural networks, in *Methods in Neuronal Modeling,* Koch, C. and Segev, I., Eds., MIT Press, Cambridge, MA, 1992, 291–333.
35. Bullock, T. H. and Diecke, F. P. J., Properties of an infra-red receptor, *J. Physiol.,* 134, 47–87, 1956.

Index

D

E

F

G

H

I

J

K

L

M

N

O

P

T

U

V

W

X

Y